ACHIEVING SCIENTIFIC LITERACY

ACHIEVING SCIENTIFIC LITERACY
From Purposes to Practices

Rodger W. Bybee

HEINEMANN
Portsmouth, NH

Heinemann
A division of Reed Elsevier Inc.
361 Hanover Street
Portsmouth, NH 03801-3912
Offices and agents throughout the world

Library of Congress Cataloging-in-Publication Data
Bybee, Rodger W.
    Achieving scientific literacy: from purposes to practices / Rodger W. Bybee.
        p.   cm.
    Includes bibliographical references.
    ISBN 0–435-07134-3
    1. Science—Study and teaching—United States. 2. Literacy—United States.
I. Title.
Q183.3.A1B92   1997
507.1'073—dc21   96-45221
                                                                            CIP

Editor: *Leigh Peake and Victoria Merecki*
Production: *J. B. Tranchemontagne*
Manufacturing: *Louise Richardson*
Cover design: *Linda Knowles*

Printed in the United States of America on acid-free paper
00 99 98 97 **ML** 1 2 3 4 5 6 7 8 9

For
Audrey Champagne
Angelo Collins
Harold Pratt
Karen Worth

who know the agony of the struggle and the message of hope.

CONTENTS

ACKNOWLEDGMENTS

Writing this book required substantial resources, support, and time. I have been extremely fortunate to have had all three when I needed them and in the forms that I needed them. Here I extend my appreciation to all who asked and listened, and read and recommended.

Although I acknowledge the contributions of colleagues, I must accept responsibility for the ideas expressed in this book. Further, the views expressed in this book do not represent official views of the National Research Council, the National Academy of Sciences, the National Academy of Engineering, the Institute of Medicine, or any constituent boards, commissions, or committees associated with these institutions.

I express my gratitude to friends and colleagues who contributed in numerous ways to the ideas presented in this book. Here, I acknowledge the discussions, arguments, and insights that developed in the course of work on the *National Science Education Standards*. For four years I worked with Audrey Champagne, State University of New York at Albany; Angelo Collins, Vanderbilt University; Harold Pratt, National Research Council; and Karen Worth, Education Development Center. Although we argued, I am convinced the balance of perspectives within this group provided a strong and appropriate set of national standards. Because of their understanding and work on the goal of achieving scientific literacy, I dedicate this book to them.

This book literally grew out of an earlier work, *Reforming Science Education: Personal and Social Reflections* (Bybee, 1993). In that book I presented ten essays published between 1977 and 1993. At the request of Brian Ellerbeck, executive editor at Teachers College Press, and Richard Duschl, editor for the series and professor at the Vanderbilt University, I began preparing a final section for that book in which I described and analyzed the contemporary reform of science education. I soon realized that the final section had a tone and style that markedly contrasted with the rest of the book. I was creating a new book, one that generally described the reform of science education and specifically addressed the issue of achieving scientific literacy. Fortunately, Brian and Rick both understood and encouraged my work, and I thank them for their insights, encouragement, and support.

Other individuals have directly contributed to ideas in this book. Decker Walker, Stanford University, provided insights about the various dimensions of educational reform; Douglas Roberts, Calgary University, continues to teach me about the science curriculum; Gordon Uno, Oklahoma State University, worked closely with me on *Developing Biological Literacy* (Biological Sciences Curriculum Study, 1993), the result of which was a clarification of my ideas about scientific literacy; Elizabeth Stage, University of California, always listened and provided insights and suggestions; David Heil, past associate director of the Oregon Museum of Science and Industry, never let me forget the role and importance of informal education; and Susan Loucks-Horsley, WestEd and the National Research Council, raised my awareness about the concerns of teachers and the essential role of professional development. She also made insightful recommendations in the final stages of preparing this book.

For over a decade, I worked with Joe McInerney, director of the Biological Sciences Curriculum Study. Joe has provided numerous lessons, but the one that stands out is the essential link between clear thinking and good writing. I have benefited from discussions with Janet Carlson Powell, who always seemed to ask questions about those things that I had least thought about. Other Biological Sciences Curriculum Study staff who contributed include Jim Ellis, Don Maxwell, Gail Foster, Nancy Landes, Lynda Micikas, and Mike Dougherty.

In the final stages of completing this book, I accepted a position as executive director of the Center for Science, Mathematics, and Engineering Education at the National Research Council. While at the National Research Council, I have enjoyed the support of Bruce Alberts, president of the National Academy of Sciences, and Donald Kennedy, president emeritus of Stanford University and chair of the Center's Advisory Board. Bruce and Don exemplify the leadership that we will need to achieve scientific literacy. I also have had support and benefited from discussions with other colleagues at the National Research Council: Joan Ferrini-Mundy, Danny Goroff, Nancy Devino, Jan Tuomi, Patrice Legro, Diane Mann, Maureen Shiflett, Sandy Wigdor, Jeanette Offenbacher, and Kirsten Sampson.

C. Yvonne Wise provided more than her computing and word-processing skills. She pointed out errors, identified omissions, and improved my writing through her understanding of grammar and style. In the same vein, Leigh Peake's recommendations improved the narrative and the overall structure of the book. Thank you to Linda Howe and Alice Cheyer for their careful editing and to Joanne Tranchemontagne for expertly guiding this manuscript through production.

Finally, my wife, Patricia, listened to me explain ideas from the book, read and commented on chapters, and always provided support for my work.

PROLOGUE

Achieving scientific literacy—in this endeavor, are we sinking, drifting, or sailing? A nautical metaphor provides some useful insights. If science education is sinking, what would this mean? Perhaps we would have reports of the United States moving downward on international assessments. We would probably have articles on science programs passing into weaker and worsened conditions. Some might say that the value of a science education has declined and that our goal of scientific literacy was an unrealized and unrealizable myth. Others would simply declare that our efforts to reform have failed. One can certainly find some examples of such statements. But I do not think the evidence suggests that science education is sinking. We have not reached the point where we know all is lost and that we begin emergency procedures to abandon our educational ship. The tremendous activity in the science education community at the local, state, and national levels does not seem to be focused on survival. Rather, it seems to be directed toward repair and improvement, and some progress toward a destination. I conclude, at least temporarily, that science education is not sinking.

Are we drifting? Is the reform of science education being carried by currents and winds? Are we proceeding without resistance toward an uncertain destination? To me, this sounds more like our situation today. For about two decades, science education has drifted into the future. For brief periods, various pedagogical winds and political currents have determined the direction of educational change. If we are drifting, we can relax and enjoy the trip unless, of course, we are concerned about reaching a specific destination, such as achieving higher levels of scientific literacy for all students. In some respects, science educators at all levels, kindergarten through college, have enjoyed a trip filled with reports, research, and rhetoric about current problems. However, too many have confused fads with reform and jargon with improvement of science education. Do you remember teaching machines, individualized instruction, behavioral objectives, and other fads that were painfully instituted and then discarded because they did not solve the problems of science education?

Reforming science education and achieving scientific literacy for all students should not be confused with a convention workshop, a summer institute, a district science committee meeting, or a textbook adoption. Reforming science education and achieving scientific literacy is not a make-it and take-it workshop; it is not a lesson to use on Monday; it is not a new widget, gadget, or gimmick; it is not a new computer, electrophoresis gel chamber, or pH meter; it is not an inservice day on the latest instructional technique or assessment strategy; it is not writing a report, doing esoteric research, or introducing future science teachers to curriculum projects from the 1960s. Certainly, all of these may be helpful, sometimes insightful, and occasionally motivational, but they are winds and currents in science education. Equating such activities with the need for long-term sustained reform is confusing a drifting path with a determined movement toward a definite destination. I would argue we have been drifting when we should have been sailing.

This book is about sailing. The time has come for the science education community to set our sights, stop drifting, and begin sailing. What might this mean? The metaphor of sailing suggests several things. First, sailing implies a destination or goal, that we are embarking on a voyage, and that science educators are taking the helm, managing the supplies, and changing the sails. We are assuming responsibility for steering toward our destination and some control for the journey. I selected the metaphor of sailing rather than, say, rowing because it suggests that each individual's work can be assisted by the skillful use of outside forces such as winds and currents. In sailing, the ability to tack is crucial. Tacking implies changing course. It suggests taking a course of action that minimizes opposition and attains a goal. What appears to be a wind against you can become a force that assists you, if you know how to steer your course. I maintain there are ample winds of change, and indeed some appear to be blowing against science educators. We must learn how to tack into those winds and use them to our advantage. Second, in sailing there is a continuous need for navigation, for determining the present location, evaluating the weather, assessing stocks of supplies and materials, and making the necessary midcourse corrections in order to arrive eventually at the designated port. Finally, the metaphor of sailing implies that we approach the goal of achieving scientific literacy with vigor and energy, that we sail into the improvement of our policies, programs, and practices. So, concerning the goal of scientific literacy, we are not sinking, many are drifting, and some are sailing. In this book, I argue that we need to increase the number of those who are sailing.

The challenge of achieving scientific literacy requires all of us to set our sights on larger goals and prepare for a longer journey. Several critical, but seldom discussed, activities can help us do so. First, there should be a thoughtful consideration of the history of science education and what the combined efforts of the community of science educators can reasonably contribute to the development of a scientifically literate citizenry. Many individuals do not want to consider such issues, and I maintain that is a major part of the problem. Second, sailing requires a considered dialogue about documents that define our destination, namely the *Standards and Benchmarks*. I do not mean an academic analysis of the similarities and differences between these documents; I mean thoughtful debate and interpretation of the meaning of the *Standards* and *Benchmarks* by the individuals who have to design new programs and implement new classroom practices. The aim of this dialogue is learning itself, not the practical solution to everyday problems. Each individual must confront the unique problems of his or her situation. Hope lies in the consistent translation of fundamental ideas into the unique forms that individuals encounter. Third, we must think of science education both as a system and as the components of a system. This means, for example, approaching reform systematically and not focusing on specific components, such as lessons, topics, curriculum, assessment, and teacher education. Fourth, whether local, state, or national, science educators have to construct a vision and develop a plan for achieving scientific literacy. The vision must be informed by contemporary thinking about the purposes of science education, student learning, and what we can expect from curriculum programs and school personnel. Fifth, science educators have to develop an understanding of educational change and actively apply this understanding to the implementation of new science programs and teaching practices. Finally, an essential point, all of us have to internalize the idea that *we are the reform*. Reforming science education and achieving scientific literacy are not "out there" in reports, frameworks, and assessments; they are what we do in our daily work as members of the science education community. The community will be well served by recognizing the roles and contributions of all members and the need to begin cooperating. We must set sail toward our collective goal of achieving higher levels of scientifically literate citizens.

The following brief personal note introduces the experiences and context of my journey to the positions and ideas in this book. Between 1985 and 1995, I worked at the Biological Sciences Curriculum Study and with support from the National Science Foundation had the opportunity to direct several major curriculum projects: a program for elementary schools—*Science for Life and Living: Integrating Science, Technology, and Health;* a program for middle schools—*Middle School Science and Technology;* a program for high schools—*Biology Science: A Human Approach;* and an introductory biology program for undergraduates—*A New Model for Introductory Biology at Two-Year and Community Colleges.* We based each of these Biological Sciences Curriculum Study programs on a design study, a year-long effort to conceptualize and describe the requirements and specifications for the respective program. My work as a curriculum developer provided a perspective that attends to a national agenda and the mandates of state and local education. Developing and implementing the Biological Sciences Curriculum Study programs required attention to national reports, state guidelines, and local initiatives, and to classroom requirements through field testing by hundreds of science teachers and many thousands of students.

Beginning in 1992, I also worked on the *National Science Education Standards* project, first as a member of the content working group and then, between 1993 and 1995, as chair of that group. This experience gave me a new perspective on educational reform, namely, the importance of policies, their contribution to a definition of scientific literacy, and their role in guiding reform. I also clearly recognize the fact that national, state, or local standards are not science programs.

There is a critical need to understand standards and to make the translation of standards—as policies—into science programs and instructional practices.

In September 1995, I left the Biological Sciences Curriculum Study and joined the National Research Council as executive director of the Center for Science, Mathematics, and Engineering Education. This position reinforced my original ideas about the personal need for a national as well as state and local perspectives.

In this book, I have responded to several critical issues relative to contemporary reform. Although the rhetoric for reform is national in scale and tone, the reality of reform is occurring at the state and local levels. A second issue extends from the first; namely, reforming science education requires that science educators at the local level acquire a better general understanding of science education. By this I mean the history and nature of reform in science education, a clearer understanding of scientific literacy, and basic ideas about curriculum, instruction, assessment, professional development, and the educational system. I also think science educators at the national level must develop greater understanding of factors that influence reform at the local level. And all individuals interested in improving school science programs need a clear perspective on the specific processes, dimensions, and dynamics of reforming science education.

This book addresses a broad range of questions, issues, and needs of those who are engaged in the process of improving school science programs. In writing it I tried to keep the widest possible audience in mind—the science education community. This includes classroom teachers, district supervisors, administrators, state science supervisors, policy makers, curriculum developers, assessment specialists, scientists, engineers, health care professionals, informal educators, and parents. Specifically, I kept school personnel in mind as I wrote. Classroom teachers, science supervisors, and administrators who are directly or indirectly responsible for improving science education will find this book helpful. The book should help them with both the larger perspective of reform and with the practical issues of improving their programs and practices. Science educators who are directly responsible for initiating and coordinating reform will also find this book helpful. I am referring to state and local science supervisors, teacher educators, and scientists and engineers from business and industry, and two- and four-year colleges and universities, who are contributing to local reform through various alliances. Finally, undergraduates—the future science educators—should be aware of the perspectives outlined in this book.

This volume includes historical, philosophical, psychological, and social discussions. In my opinion, these will empower those who must ultimately understand their chosen discipline of science education, the reform, and the dynamics that will improve their unique situation. I have tried to present a broad picture for the science education community. The subtitle, *From Purposes to Practices,* points out the need to translate the purpose of achieving scientific literacy to the policies, programs, and practices that have meaning for individual science educators. Moving from purposes to practices should provide consistency and coherence for the science education system. I mention this because I wish to note that this approach is not top-down. The relationships among purpose-policy-program-practice could as well be thought of as originating at the center and moving out, or interaction and coordination among components of one system that must function with some harmony. Personally, I counter any "top-down" criticism by espousing systemic reform that focuses on standards and aims to achieve higher levels of scientific literacy for all students. Where the reform is initiated is not the issue. Clarifying our common destination, identifying what we need to sail toward that destination, and assuming responsibility for our respective roles in the journey are the issues.

Achieving Scientific Literacy: From Purposes to Practices extends and elaborates several contemporary themes: a clarification of scientific literacy, a systemic perspective on reform, a standards-based approach to improving science education, and emphasis on science teaching and student learning. In the first chapter, "Looking at the Past Fifty Years," I examine educational reform since World War II. By placing science education in the larger context of education and society and providing some insights about the "Golden Age," I provide a historical perspective on contemporary reform.

The second chapter, "Contemporary Reform of Science Education," introduces the Four P's—purpose, policy, program, practice—to analyze the dimensions, difficulties, and dynamics of reforming science education. I make several points, including the need to recognize the importance of these Four P's, the roles that different individuals play in forming purpose, policy, program, and practice, and the difficulties scaling up reform for these dimensions. Finally, the most essential and most difficult aspect of reforming science education involves changing classroom practices.

The third chapter, "Searching for Scientific Literacy," presents a historical review and development of science educators' use of the term *scientific literacy*. From the introduction of the term to our contemporary use, scientific literacy has had a fairly stable set of defining characteristics. It expresses those aspects of science that individuals should know, value, and experience. This chapter provides an introduction to different points of view relative to achieving scientific literacy. It also develops my position, one that counters the view presented by Morris Shamos in *The Myth of Scientific Literacy* (1995). Shamos argues that scientific literacy is a false and unachievable aim and that educators have been, and are being, led astray by pursuing this goal.

In Chapter 4, "Defining Scientific Literacy," I conclude that scientific literacy is a metaphor for the purposes of science education and that it expresses several factors, including a general education orientation, a continuum of understandings of and about science and technology, and a variety of dimensions of science education. I present a framework for scientific literacy that can be considered in the design of school science programs. In brief, the framework proposes that we consider different dimensions, including nominal, functional, conceptual, procedural, and multidimensional, in scientific literacy.

Chapter 5, "Establishing National Standards," addresses the history and role of standards in science education. The 1990s will certainly be recorded as one in which standards at all levels—local, state, national—guided educational reform.

Standards in science education serve several functions. They can be a measure of comparison of scientific literacy, outcomes for scientific literacy, and a vision of scientific literacy. The chapter presents standards as policies because they represent courses of action, they influence decisions, they embody guiding principles, and they define procedures for science programs and practices. The chapter focuses on the *National Science Education Standards* (National Research Council, 1996) and includes general principles applicable for states and local school districts.

Chapter 6, "Creating a Vision of Scientific Literacy," combines the theme of scientific literacy with the national standards in an effort to show the coherence and consistency of the translation from purpose to policy. I return to the framework presented in Chapter 4 and show how the national standards align with that framework.

Chapter 7, "Enriching the Science Curriculum," addresses the design of science programs. This chapter, along with the next on instruction, provides guides and suggestions for those responsible for school science programs. The chapter does not suggest a curriculum; rather it provides helpful suggestions and recommended specifications for designing a school science program.

Chapter 8, "Improving Instruction," presents an instructional model useful at the most practical level, the science classroom. The instructional model includes five phases—engagement, exploration, explanation, elaboration, and evaluation. This 5 Es model has been used successfully in several contemporary science programs at the Biological Sciences Curriculum Study, and it has been introduced by other individuals as an effective means to organize science instruction. The chapter also includes the background on other instructional models and proposes design requirements for instructional models.

In Chapter 9, "Implementing a Standards-Based Systemic Reform," I return to a broad perspective and discuss issues related to standards, large-scale reform, and systemic initiatives.

The final chapter, "The Year 2000 and Beyond," discusses my observations, reflections, and recommendations on achieving scientific literacy and the contemporary reform of science education.

The book presents a broad vision for improving science education. At times it is historical and philosophical, and at other times it is concrete and practical. I address a number of issues that we should attend to as we begin to sail toward a worthwhile destination, achieving scientific literacy for all students.

1

Looking at the Past Fifty Years

istorical perspective can provide valuable insights and teach useful lessons. However, science educators and the science education literature give little recognition to the role of history and the development of the discipline. Certainly, exceptions exist, for example, Paul DeHart Hurd's (1962) *Biology Education in American Schools: 1890–1960* and George DeBoer's (1991) *A History of Ideas in Science Education,* yet the assertion stands: the literature does not generally include historical perspectives. In this chapter, I draw on history to clarify several points that may be applied to the contemporary reform of science education (Tyack and Cuban, 1995). At this point, I should like to acknowledge Diane Ravitch's (1983) excellent *The Troubled Crusade: American Education 1945–1980.*

Overall, how would I characterize educational reform during the past fifty years? First, depending on whether you view educational reform as a gradual, evolutionary process or one of "punctuated equilibrium," we have either been involved in continuous reform since the end of World War II or we are now experiencing the third reform since 1945. Second, over the past fifty years, the call for reform and improvement has broadened to unite more diverse groups and incorporate more components of the educational system. Cross-cultural comparisons, in particular, international assessment scores in mathematics and science, support recent reforms. Third, they also have a systemic perspective, having expanded beyond different levels of education (for example, secondary) or disciplines (for example, the sciences) to include the entire educational system. Finally, the past fifty years have produced greater reform of science educational policies and programs and less of classroom practices.

Throughout this book, I use the terms *purpose, policy, program,* and *practice* to identify different essential and interdependent aspects of the science education system. Purpose refers to various goal statements, policies are plans or guides for school programs for which curriculum materials serve as an example, and practice refers to the various actions and processes of teaching. (I discuss these terms in greater detail in the next chapter.)

Historical Perspectives

As World War II concluded, Americans turned their attention to domestic issues, including education. They found a system in need of major improvement: new buildings and classrooms, because renovations had been delayed during the Depression and the war and the postwar "baby boom" would soon overcrowd extant facilities, and reinvigoration to counter the teacher shortage and low salaries. Other issues lurked beneath the surface. World War II brought America into an age dominated by science and technology, and its citizens would require higher levels of education to sustain a technologically oriented society and economic progress. More important, Americans were keenly aware that we fought a major war for the ideals of democracy, the rights of freedom, and the American way of life. With victory abroad, many Americans asked how these democratic ideals were being implemented at home. Educators immediately realized that there was a real need to address the inequalities in society, and this meant providing educational opportunity for all students.

Education After World War II

Several reports of this period presented educational policies and proposals that would influence science education in the coming decades. This discussion concentrates on the 1944 report of the Educational Policies Commission, *Education for All American Youth,* the 1945 report of the Harvard Committee, *General Education in a Free Society,* and the report that established the National Science Foundation (NSF), *Science: The Endless Frontier* (Bush, 1945). These documents provided a baseline for reforms in the 1950s, 1970s, and 1990s, and one could easily argue that they represented a reform in and of themselves.

The 1944 report *Education for All American Youth* revised and restated *The Cardinal Principles of Secondary Education,* originally published in 1918. These principles were (1) health, (2) command of fundamental processes, (3) worthy home membership, (4) vocation, (5) citizenship, (6) worthy use of leisure, and (7) ethical character. The 1944 report supported progressive ideals and the concept of general education, since it emphasized developing personal and social goals in areas such as health, leisure, family life, vocation, consumer ethics, economics, and democratic citizenship. There was, to use a phrase from the report, "no aristocracy of 'subjects'" (p. 142). Science was on a par with other school subjects.

One of the most prominent postwar reports was the Harvard "redbook," *General Education in a Free Society* (Harvard Committee, 1945). As the title suggests, the report emphasized general education, which it contrasts with special education. "It [general education] is used to indicate that part of a student's whole education which looks first of all to his life as a responsible human being and citizen; while the term, special education, indicates that part which looks to the student's competence in some occupation" (p. 51). Although the theme of general education predominated, the committee recommended academic disciplines—humanities, social studies, science, and mathematics—in the secondary school curriculum and supported ability grouping. The report also suggested that there must be "courses of different difficulty and different methods in each of the three spheres of general education" (p. 100). The criterion for enrollment in the courses was ability. In his introduction, James Bryant Conant, the president of Harvard, presented ideas on general education and ability that anticipate his position in several subsequent reports, which stimulated reform in the late 1950s and significantly influenced the Golden Age of science education. The emphasis on general education fore-shadows its importance as a central orientation in scientific literacy and in documents such as the *National Science Education Standards* (National Research Council, 1996).

General Education in a Free Society contained an excellent discussion of the place and role of science and technology in general education.

> Science education in general education should be characterized mainly by broad integrative elements—the comparison of scientific with other modes of thought, the comparison and contrast of the individual sciences with one another, the relations of science with its own past and with general human history, and of science with problems of human society. These are areas in which science can make a lasting contribution to the general education of all students. Unfortunately, these areas are slighted most often in modern teaching. (p. 155)

This passage sounds much like contemporary statements from *Science for All Americans* (Rutherford and Ahlgren, 1989), those advocating the Science Technology Science (STS) theme (Bybee, 1987; Yager, 1993), and those recommending teaching more about the history and nature of science and technology (Bybee et al., 1991). Immediately following this section of the text, the Harvard Committee made another insightful statement.

> Many science teachers may at once object that they are already badly pressed for time. There is too much ground to cover, and so much more is added day by day, that the teacher is engaged in a continuous struggle to encompass the subject matter. How is he, then, to deal with extra things—the critical examination, history, literature, and general cultural context of his subject? It is of course true that as extra things these aspects of science instruction should be impossible. But they are not extra things—they are the very stuff of science in general education. (p. 156)

Clearly, science education is a component of general education. Although the idea of general education has not been realized to any significant degree in science programs, it continues to emerge as an important, if not essential, orientation in contemporary discussions of scientific literacy.

Vannevar Bush's (1945) report *Science: The Endless Frontier* called for the establishment of the NSF and influenced the subsequent relationship of NSF to reforms in science education. In 1944, President Franklin Roosevelt had asked Bush to prepare a report on the role of science and technology in times of peace. *Science: The Endless Frontier* made a strong case for the importance of science in society. It also established a primary connection between science education and the emergence of new talent in science and engineering, and recommended scholarships and fellowships for undergraduates. About the secondary connection, as most science educators recognize, the report concluded that "Improvement in the teaching of science is imperative, for students of latent scientific ability are particularly vulnerable to high school teaching which fails to awaken interest or to provide adequate instruction" (p. 26). In contrast to the other reports of this period, this statement suggests that the entry of NSF into curriculum development was prompted by the goal of specialized education and the need to enlarge the pool of men and women qualified for advanced study in science and engineering. Congress passed the National Science Foundation Act in 1950.

James Bryant Conant served as a member of the Discovery and Development of Scientific Talent Committee and directly influenced the policies on science education as they were expressed in the final report, which also quoted his position: "in every section of the entire area where the word science may properly be applied, the limiting factor is a human one. We shall have rapid or slow advance in this direction or in that depending on the number of really first-class men who are engaged in the work in question. . . . So in the last analysis, the future of science in this country will be determined by our basic educational policy" (p. 23). Conant clearly emphasizes the need for talented students of high ability who might eventually pursue careers in science and engineering. His influence on science education reform in the late 1950s and 1960s was probably much greater than science educators have either recognized or acknowledged.

After World War II, several major educational issues moved into the national spotlight. Progressivism was evident in the schools, but to the extent that progressive education still had influence, reports such as *Education for All American Youth* and *General Education in a Free Society* had transformed its basic tenets to recognize the academic disciplines within general education. Education became a primary means for social change. Educational questions did not center on curriculum or instruction; rather, they addressed larger social problems such as civil rights. Science education received due recognition, given that society began to acknowledge the place and importance of science and technology, but for the most part, it was discussed within the context of general education.

Lawrence Cremin (1988) ended *American Education: The Metropolitan Experience* with the observation that popularization was an abiding characteristic of education in America. Popularization means making education available and accessible to increasingly diverse groups and individuals and touches on issues such as the rights of minorities, racial segregation, busing as a means of integration, the legality of school prayer, and mainstreaming disabled students. Cremin returned to the theme of popularization in *Popular Education and Its Discontents* (1990), completed just before his death. The expectation that the school should fulfill social aspirations has led to at least three identifiable postwar reforms.

Science Education After World War II

Just before the war, science educators had prepared *Science in General Education* (Commission on Secondary School Curriculum, 1937), a report that addressed such essential issues as the relation of science teaching to general education and the needs of adolescents in the areas of personal living, personal-social relations, social-civic relationships, and economic relationships. Although the report provided excellent lists of generalizations and teaching suggestions in these areas, the war overshadowed its importance, and after the war, the times demanded a different, more functional orientation for science education.

The forty-sixth yearbook of a report produced by the National Society for the Study of Education (NSSE) (1947), *Science Education in American Schools,* represents a clear statement of the postwar approach to science education. After an opening chapter on "Science and People," the committee turned to "Objectives of Science Teaching": "Science concepts and principles must also be taught so that they will be *functional*" (p. 26). The report continues with a section that could have been written today instead of fifty years ago.

> The critical element in functional concepts and principles, as in functional information, is understanding. Understanding is not quickly achieved. It rarely results in any useful amount from a single experience or from exact duplicates of that experience, however often it is repeated . . . For the kinds of concepts and principles which are properly science objectives there must be many and varied experiences in which the same idea, large or small, occurs in differing situations. Moreover, for the most fruitful learning, these experiences must be arranged and graded with respect to complexity and difficulty, so that the pupil may be guided to organize his meanings at higher and higher levels. Meaning learning is spiral. Each experience adds a new loop in the spiral of meaning. (p. 27)

These words suggest a close connection between curriculum and instruction and hint at a general model of teaching. The report went on to mention a range of objectives for science teaching:

- functional information or facts (for example, living things consist of plants and animals)

- functional concepts (all life evolved from simple forms)

- functional understanding of principles (all living things reproduce their own kind)

- instrumental skills (using simple equipment)

- problem-solving skills (sensing a problem)

- attitudes (open-mindedness)

- applications (scientific contributions)
- interest (hobbies)

The First Educational Reform: Attracting Academic Talent

Following World War II, many school programs embodied ideas from the progressive education movement. One, that schools should retain larger percentages of students into higher grade levels, resulted in a proliferation of optional courses, such as vocational education, civics, health, and recreation. A second progressive idea, that school programs should be interesting to students and teachable, prompted a shift to the practical: arithmetic became consumer arithmetic and science became utilitarian science (Cremin, 1990). Education emphasized life adjustment. Whatever the name, in the late 1940s educational policies clearly recommended a functional approach to instruction using everyday situations and objects. Classes and lessons should change students' behavior and attitudes so they would be well adjusted, conform to society's norms, and live effective lives as workers, family members, and citizens (Ravitch, 1983).

The academic community and many conservatives criticized "life adjustment," or functional education, claiming that the progressive movement had gone too far. Indeed, it had lost support and could not sustain such criticism. Then, on October 4, 1957, the Russians successfully orbited *Sputnik I,* and the nation's attention immediately turned to science and mathematics, foreign languages, higher academic standards, and a search for high-ability students. The 1950s had also witnessed loyalty investigations and searches for subversives and communists on the one hand, and movements for racial equality and educational opportunity on the other: in 1954, the Senate voted to censure Joseph McCarthy and the Supreme Court ruled on the *Brown v. Board of Education* case.

James Bryant Conant

The work of James Bryant Conant stimulated the first postwar educational reform. In 1959, he published *The American High School Today,* in 1961, *Slums and Suburbs,* and in 1964, *Shaping Educational Policy.* For Conant, the goals of education were both social and intellectual. Thus, comprehensive high schools enrolling diverse groups of students are important to the cohesiveness of society and the "melting pot" ideal. However, Conant suggested, in the course of achieving this social goal, schools should not sacrifice the quality of education they offered to the academically talented. "Can a school at one and the same time provide a good general education for *all* the pupils and future citizens of a democracy, provide elective programs for the majority to develop useful skills, and educate adequately those with a talent for handling advanced academic subjects—particularly foreign languages and advanced mathematics?" (p. 15). To this day, we continue to struggle with this question. Most recently, the *National Science Education Standards* engaged the issue of equity and excellence, a cultural paradox if not a contradiction. The question remains: how can we design programs so that all students have equal opportunity and some receive advanced education according to their individual potential?

Conant considered how to preserve the quality of education for the academically talented, which necessarily required rigorous academic standards, knowing that there was inevitable erosion of those standards when applied to diverse student populations. As a remedy, he proposed early classification and tracking, combined with comprehensive basic courses for all students. His research had identified eight schools in this country that were satisfactorily fulfilling the three main objectives of a comprehensive high school. He also encountered a previously unacknowledged inequity. In these same schools, he reported, "the academic inventory showed that more than half the academically talented boys had studied at least seven years of mathematics and science as well as seven years of English and social studies. . . . On the other hand, in no school had a majority of the academically talented girls studied as much as seven years of mathematics and science" (1959, p. 22).

Conant thus brought the issue of education for girls to public attention. Later, in *Slums and Suburbs* (1961), he considered the disadvantages of being a minority in American schools, especially if those schools were located in large urban areas.

Conant recommended that

> All students should obtain some understanding of the nature of science and the scientific approach by a required course in physical science or biology. This course should be given in at least three sections grouped by ability. . . . To accommodate students who are interested in science but do not have the required mathematical ability, two types of chemistry courses should be offered. For entry into one, at least a C in algebra and tenth-grade mathematics should be required. The other course should be designed for pupils with less mathematical ability. The standards even in this second course, however, should be such that those with less than average ability (assuming a distribution of ability according to the national norm) will have difficulty passing the course.
>
> In addition to the physics course given in the twelfth grade with mathematics as a prerequisite . . . , another course in physics should be offered with some such designation as "practical physics." The standards in this second course should be such that students with less than average ability have difficulty passing the course. (p. 73)

This recommendation presents what Conant considers a good general education in science for all pupils that also offers opportunities for advanced scientific study to those with the requisite talent and skills. Clearly, the courses are different (and Conant does not address the issue of science courses for students with less-than-average ability who have difficulty passing). Many high schools implemented Conant's recommendations only to find that the majority of students elected not to take any science course after tenth-grade biology (Hurd et al., 1981).

Providing a curriculum for all students while also accommodating academically talented individuals remains an issue in contemporary education. In the end, Conant opted to stop the erosion of academic standards within the basic curriculum, and to identify students with academic potential early and send them on a different track.

James Bryant Conant's reform efforts did not stop with these recommendations for the American high school. In 1963 he published *The Education of American Teachers,* a report based on visits to seventy-seven institutions in twenty-two states and discussions with hundreds of professors and teachers. Conant recommended drastic changes in the education of teachers; in fact, he would have revamped the entire system. Among the more significant changes, he recommended

- moving the responsibility of teacher certification from states to colleges and universities

- establishing actual performance in the classroom as the major criterion for certification

- shifting a significant portion of teacher education to the schools

- replacing teaching methods classes with experience in classrooms and supervision by teachers

Conant's assessment, that teacher education required radical reform, left few with a sense of complacency. Eliminate the state certification shibboleth and establish quality control centers in colleges and universities, school systems, and state departments of education, each with its own unique contribution. *The Education of American Teachers* initiated a national debate that included the academic community, school personnel, and the public. Much of what Conant talked about still echoes in the meeting rooms of local schools where committees are discussing the contemporary improvement of science education. Unfortunately, one seldom hears the name James Bryant Conant mentioned.

Conant concluded his 1959 report on high schools on this note:

> I am convinced American secondary education can be made satisfactory without any radical changes in the basic pattern. This can only be done, however, if citizens in many localities display sufficient interest in their schools and are willing to support them. The improvements must come school by school and be made with due regard for the nature of the community . . . (p. 96)

Although he emphasizes school-based reform, he clearly recognizes the importance of diversity and a regard for the community. These insights are as important today as they were in 1959.

I suggest that James Bryant Conant's ideas and policies had a much greater impact in the 1960s than has previously been acknowledged. Through *General Education in a Free Society* (Harvard Committee, 1945), his contribution to *Science: The Endless Frontier* (Bush, 1945) and the subsequent policies at NSF, and his own research reports, Conant championed the academic disciplines. Although we have come to identify Jerome Bruner's *The Process of Education* (1960) with science curricula, I think much greater recognition is due Conant for the overall character of educational reform in the 1960s, particularly the emphasis on the academically talented in science. Significant shifts in educational policy had already occurred, mostly informed by the push to "popularize" education (Cremin, 1990). In science education, we tend to focus on curriculum-development projects and often fail to recognize the reforms already in place. In the 1960s, the science curriculum clearly dominated much of the educational reform landscape, but the influence of individuals like James Bryant Conant had done much to shape the educational system within which science educators were attempting to implement new curricula.

Programs Without Policies: The 1950s and 1960s

In 1956, Jerrold Zacharias of MIT compared education with high-fidelity recordings. You need a good performer, a good record, a good electronic system, a good room for acoustics, and a good listener. Most important of all, you need a good composition, for without a good composition everything else is pointless. For Zacharias the analogy worked, and he attempted to supply the school system with great compositions in the form of science curricula. His ideas led to the formation of the Physical Sciences Study Committee (PSSC) in 1956, the year before the launching of *Sputnik*.

A dearly held misconception of science educators claims that the "Golden Age" of science education was a response to *Sputnik*. But long before *Sputnik,* the American public had expressed discontent with the schools, and scientists and educators were concerned about the shortage of graduates in science and engineering. Critics of progressive education complained about the neglect of academic disciplines in favor of life-adjustment courses. When the Russians launched *Sputnik,* reforms were already in progress. The University of Illinois Committee on School Mathematics, for example, had begun its revision of the secondary school mathematics curriculum in 1952, and PSSC the development of a new high school physics course in 1956. What role, then, did *Sputnik* play? *Sputnik* came to symbolize the threat to America's political hegemony in the Cold War and served as a reminder, to politicians especially, of the role technology played in social progress. So powerful was this symbol that it influenced people of all political backgrounds and focused their efforts on the mission of improving education, especially in science and mathematics.

Sputnik succeeded in accelerating reform, generating public support, and increasing federal funding. The perceived superiority of the Soviet Union stimulated Congress to pass the National Defense Education Act (NDEA) in 1958, which supported student fellowships and grants to study science and mathematics, and the purchase of materials and equipment for schools. But there was an important omission. The reform during this period centered on the science curriculum and did not extend in any significant way to include policy, although some policy statements were issued. These policies were largely ignored, however, the assumption being that all would be well if the curriculum, in particular, the content, was improved.

Of the two major policy statements published at the time, one, *Rethinking Science Education,* was developed for the NSSE by a committee chaired by J. Darrell Barnard (1960). A second policy statement, *Policies for Science Education,* was prepared for the Science Manpower Project and edited by Frederick Fitzpatrick (1960). As policy statements, both reports have greater historical significance than practical influence. The critical thrust of the reform went directly from purposes to curriculum programs, bypassing the intermediary step of formulating policy. *Rethinking Science Education* attempted to integrate statements from the leading science educators of the time on diverse topics: the purposes of science education; the learning, creativity, and personalities of scientists; the status of science teaching; science in general education in colleges; improving elementary and secondary school science; curriculum development; supervision; facilities and equipment; science teacher education; professional growth; research; and problems and issues. The NSSE volume presented general policies in these and other areas, although the authors probably did not think of it as a policy document. Likewise, it gave only passing recognition to the curriculum projects that eventually dominated the 1960s.

In contrast, *Policies for Science Education* focused directly on policies, defining them as "decisions which lead to *planned programs of action*" (pp. 20–21; italics in original). This volume clarified their role and recommended programs in elementary and junior high schools, and biology and physical science programs in high schools.

The role of the NSF in science education increased dramatically after *Sputnik* and has continued, although not with criticism (Tressel, 1994). Congress had established the NSF in 1950 to support research in science and education for those entering scientific careers. Initially, NSF played no significant part in precollege science education, but as the decade progressed the role of NSF expanded steadily. In 1956, NSF supported the PSSC project and quickly expanded to curriculum projects in biology, chemistry, earth science, and the social sciences.

The "Golden Age" of science education began in the late 1950s with the development of curricula in physics, chemistry, biology, and the earth sciences. The secondary school science programs eventually became known by their acronyms: the Physical Science Study Committee became PSSC physics, the Chemical Education Materials Study became CHEM-Study and the Chemical Bond Approach Project became CBA chemistry, the Biological Sciences Curriculum Study became BSCS biology, and the Earth Science Curriculum Project became ESCP earth science. At the elementary level, the American Association for the Advancement of Science (AAAS) developed Science-A Process Approach (S-APA), Education Development Center developed Elementary Science Study (ESS), and Lawrence Hall of Science developed Science Curriculum Improvement Study (SCIS). Curriculum development continued into the 1960s and most of the new programs were published during that time. The NSF also supported a massive teacher education effort in the form of inservice and summer programs for science teachers. Generally, these teacher education programs focused on one of the new science programs and included updates on science content and new instructional strategies.

Science curriculum reformers during the "Golden Age" shared a common set of perceptions that were based in the widespread discontent with American education. First, extant school programs in the 1950s underemphasized academic disciplines; second, little support existed for academically talented and high-ability students; and third, teaching methods omitted laboratory experiences. Those who directed reform efforts thought that adequate levels of funding and knowledgeable curriculum developers could reform science education in American schools. Most of the major curriculum projects were directed by scientists and scholars with an understandable bias toward their own disciplines, the important place of knowledge, and the need for scientists and engineers, and the resulting curriculum materials were shaped by their positions. Programs were discipline-based and designed for above-average students, and supported by well-meaning social aspirations and funding priorities of NSF. These are statements of fact rather than criticism.

Another characteristic of the "Golden Age" was its focus on curricula: reform would be accomplished through the development and implementation of new science programs. Later, massive teacher retraining efforts would present the content and pedagogy associated with new curricula. According to the reformers of the 1960s, the new curriculum programs would replace extant materials and teaching methods. Their goal was not revision or reformation, it was replacement of the old with the new. In a year or so, science teachers would drop their old program and begin using the innovative new one.

Traditionally, science teaching consisted of lecture, discussion, and recitation. Science teachers relied on a single textbook, introduced few, if any, laboratory experiences, and used only an occasional film. The reformers planned to replace these materials and methods with well-illustrated, up-to-date textbooks, inquiry and discovery laboratories, and multimedia packages. The new programs would teach the structure of the scientific discipline and the modes of scientific inquiry (Bruner, 1960). The social environment of the early 1960s supported educational reform, especially in science and mathematics. There was personnel, energy, adequate funds, and scientific leadership. But these changes in science education did not accomplish as much as the reformers expected.

Implementation of the new programs was less successful than developers had hoped. A United States Office of Education survey conducted in the 1964–1965 school year found that only 20 percent of students taking introductory physics were in PSSC classes. Even more disappointing, the enrollment of high school seniors taking any physics had dropped from 25.8 percent in 1948–1949 to 19.6 percent in 1964–1965 (Watson, 1967). By the mid-1970s, the decade in which one would expect the new programs to be thoroughly implemented in school systems, a national survey found higher levels than the aforementioned results, but the percentages were still disappointing given the national effort (Weiss, 1978).

Is there an explanation for what happened? I offer several that relate to social changes in the 1960s, an external analysis, and others that center on problems perpetuated by reformers within science education, an internal analysis.

In the mid-1960s, several domestic issues refocused society's attention: violence against blacks and civil rights workers in the South; the assassination of President John F. Kennedy; the discovery of the "other" America living in poverty; and protest against the war in Vietnam. The issues of segregation and integration and the problems of the disadvantaged became educational priorities. From the pursuit of scientific superiority over the Soviet Union, the search for academically talented youth, and the emphasis on academic disciplines in schools, the priorities of government agencies and private foundations shifted to the needs of minority children and the disadvantaged. When Congress passed the Elementary and Secondary Education Act (ESEA) in 1965, the shift away from the earlier goals of improving science education was clear.

In the 1960s, science-related issues also emerged in the social arena. An example that epitomizes this shift has earned a place as a classic in the literature of the environmental movement: Rachel Carson's *Silent Spring,* published in 1962. This powerful book directed the world's attention to the detrimental effect of the indiscriminate use of man-made chemicals, which could "linger in the soil," "slow the leaping of fish," and "still the song of birds." If we continued to contaminate the environment, one day we would experience a "silent spring." Rachel Carson went beyond the evidence. She was criticized for the book's strong message and shocking rhetoric, but through her labor, the environmental movement was born. For the remaining years of the 1960s, public policy began to acknowledge the effect of human activities on the environment. In 1965, Congress passed the Clean Air Act (subsequently renewed in 1970 and 1975) and the Solid Waste Disposal Act; in 1966, the Species Conservation Act; and in 1969, the National Environmental Policy Act. In 1970, President Richard Nixon created the Environmental Protection Agency, one of the most powerful regulatory agencies in government.

Other science- and technology-related issues also influenced science education reform. The late 1960s witnessed one of the greatest technological achievements in human history. To paraphrase John F. Kennedy's goal for the decade, men landed on the moon and safely returned to the earth. Those advocating social responsibility to the poor, however, compared the amount of money allocated to space and that required to reduce poverty. Technological advances were identified with the achievements of travel in space but also with the power of destruction in war. The advantages of industrial growth were weighed against the disadvantages of pollution. By the end of the decade, some of the science-related social goals that were important at the beginning were achieved, resolved, or forgotten as entirely new problems emerged. Since these changes were for the most part external, they remained beyond the control of science education, which could, at best, give some attention to social issues, especially those related to science and technology.

A second explanation for the nonsuccess of the new programs, one which provides more in the way of insight for contemporary reform, is based on problems within science education and the perceptions of those who exercised leadership during the 1960s reform. Paradoxically, the greatest strength of the "Golden Age" also became its greatest weakness. Scholars and scientists brought the power and insights of their disciplines to the development of some of the best curriculum materials in our history—those who provided NSF institutes for science teachers presented the most up-to-date content and pedagogy—yet most of the reform efforts stopped at the schoolhouse door, and those that made it into the classroom were short-lived or largely subverted. The reformers ignored the experiences of the past, in some measure because of their negative attitude toward education and in some measure because they focused their attention on science knowledge and curriculum construction. They ignored the infrastructure of policy and programs that had to support the innovative curricula they designed and attempted to implement (Hall and Loucks, 1978). Where advocates of progressive education had ignored the academic disciplines in favor of a child-centered approach, the reformers of the 1950s and 1960s did the reverse, emphasizing the curriculum and largely ignoring other components of the educational system. An overgeneralization? Perhaps, but I would maintain that it is more accurate than not.

A second issue arises out of the perceptions and orientation of the reform leadership. Unless schools and teachers change, the curriculum, however superior, has at best only a marginal effect. To borrow a metaphor from Jerrold Zacharias, you can develop "great composition," but if you do not have support for that musical genre and someone to play the composition, the exercise is futile. The 1960s reform assumed that students would learn if the teachers taught the materials as prescribed by the curriculum developers. This issue was exacerbated by the attempt to construct "teacher-proof" materials that would work regardless of the teacher's knowledge, methods, or beliefs about teaching. Reformers ignored the power structure of schools and the possibility that science teachers would accept or reject, implement or abandon any curriculum. In addition, reformers focused on specific disciplines without attending to school science programs. Because the curriculum is more than a series of lessons and units, simply trying to replace one subject in one discipline while ignoring a school's program challenges any current system beyond its adaptable limits.

One final error, suggested by Lawrence Cremin in *The Genius of American Education* (1965), was that reformers failed to ask the questions that are most central to education: What is the purpose of education? What kind of individuals and what kind of society do we hope to achieve through education? What kind of curriculum and instruction are required to attain these goals? These questions can be synthesized into the Sisyphean question in science education: What should the scientifically and technologically literate person know, value, and do—as a citizen?

The Second Educational Reform: A Crisis in the Classroom

Beginning in the late 1960s the American public witnessed a flow of popular books criticizing the schools, for example, Jonathan Kozol's *Death at an Early Age* (1967), Herbert Kohl's *36 Children* (1967), Nat Hentoff's *Our Children Are Dying* (1966), and John Holt's *How Children Fail* (1982). In 1970, Charles Silberman published *Crisis in the Classroom,* a well-researched and documented critique of American education that generally confirmed contemporary criticism. According to Silberman, the central problem of American schools was mindlessness. Although school personnel are decent, intelligent, and caring, they do not ask *why* they are doing what they are doing. Silberman clarifies his indictment: "This mindlessness—the failure or refusal to think seriously about educational purposes, the reluctance to question established practices—is not the monopoly of the public school; it is diffused remarkably evenly throughout the entire educational system, and indeed the entire society." He goes further and proposes a solution: "If mindlessness is the central problem, the solution must lie in infusing the various educating institutions with purpose, more important, with thought about purpose, and about the ways in which techniques, content, and organization fulfill or alter purpose" (p. 11).

Silberman's indictment opens a window to the soul of educational reform, and, in large measure, the situation has not changed. We still do not ask the *why* questions, concentrating instead on answering the *how* questions. For example, we spend much more time and effort on how-to activities than we do on thoughtful consideration of the larger purposes of science education and science teaching. What I am suggesting, lest my statements be misinterpreted, is that reforming science education will not be accomplished simply by talking and writing about educational purposes. There is little question about the need for thoughtful discussions at local, state, and national levels as well as among those who are responsible for educating future teachers, supervising professional development programs, and teaching students. Then we can engage in the process of translating that purpose into policies, programs, and practices that connect the philosophical and the practical.

The early 1970s also witnessed the appearance of national reports on education. The National Commission on the Reform of Secondary Education published *The Reform of Secondary Education* (Brown, 1973); the Panel on Youth of the President's Science Advisory Committee released *Youth: Transition to Adulthood* (Coleman, 1974); and the National Panel on High School and Adolescent Education released *The Education of Adolescents* (Martin, 1974). *The New Secondary Education* (Gibbons, 1976) published by Phi Delta Kappa, summarized the earlier reports and provided further perspective on reform.

The serious, crisis-oriented tone of these reports is markedly different from that of the Conant reports. All three shared a common focus: "The crucial issue of secondary education, and perhaps of all public education, is how to promote the successful transition of youth from childhood and school to adulthood and the community" (Gibbons, 1976, p. 1). All three also presented more or less parallel recommendations. They portrayed American high schools as large, unresponsive, and isolated institutions that separated young people by age group and ability and focused on academic programs. The adolescents were, in a word, alienated. They lacked the usual educational experiences that facilitated the transition from youth to adulthood. (See Table 1.1 for a summary of these recommendations.)

In general, the reports recommended more flexibility, a reduction in size, more options in the curriculum, some opportunities for experiences away from the school, and experiences that empowered adolescents to make decisions about their careers. The reports also made recommendations about both compulsory attendance laws and minimum wage laws to allow youth to get away from the high school and into the world of work.

In the reform of the 1970s, the "popularization" of education (Cremin, 1990) assumed a new character, one unfortunately less than sympathetic to science education. This second wave of reform followed on the heels of the "Golden Age" of science education, when many new science programs were introduced to help insure victory in the Cold War. Schools had the burden of teaching more science and mathematics but using materials designed primarily for the academically talented. Growing protest against an unpopular war and the expansion of numerous other personal freedoms formed a backdrop to educational debate on the role and effect of schooling in relation to equality and opportunity (Coleman, 1966). In this social context, it is not surprising that the reports of the period focused on adolescents instead of educational programs, and seldom mentioned science and mathematics. Educators attempted to involve high school students in life choices. Viewed collectively, their recommendations were designed to reconnect adolescents with the society in which they lived and worked. In the end, however, these reports had very little influence on educational programs and practices and quietly dropped from view.

During the 1970s past values and ideas about social and economic growth were called into question. Technology, in the form of the supersonic transport (SST), was challenged. After a long congressional battle, support for the SST was ended. At the same time, the public was reading books with titles like *The Limits to Growth* (Meadows et al., 1972) and *Only One Earth* (Ward and Dubos, 1972). The Organization of Petroleum Exporting Countries (OPEC) brought the message of these books home to Americans through the oil embargo of 1973–1974, which called attention to the issue of energy supplies and national vulnerability.

During the decade Congress passed a number of environmental bills, including the Water Pollution Control Act (1972), the Endangered Species Act (1973), the Toxic Substances Control Act (1976), and the Clean Air and Clean Water Acts (1977), which emphasized the interrelationship of science, technology, and society. The near disaster at the Three Mile Island nuclear reactor in 1979 came to symbolize the American public's ambivalence about science, which had been developing over two decades. The hope for cheap energy collided with a mistrust of technology, the need for energy collided with a questioning of the safety of nuclear power, and the possibility of unlimited energy collided with a profound sense of vulnerability. Public enthusiasm and support for science education funding declined. Programs from the "Golden Age" became associated with negative aspects of the schools, yet science educators found it difficult to initiate reforms. Science- and technology-related social issues increasingly gained public attention and some made negative connections between these issues and the need for better science education.

TABLE 1.1 *Summary of Recommendations*

The Reform of Secondary Education (Brown, 1973)	Youth: Transition to Adulthood (Coleman, 1974)	The Education of Adolescents (Martin, 1974)
1. *Goals:* Every school and community should develop new goals for the schools.	1. *Change School Structure:* Reverse trend toward comprehensive high schools, and emphasize teaching and student choice.	1. *Community:* Extend educational opportunities into the community. There should be real jobs for adolescents.
2. *Emphasis:* Curriculum should center on students' needs and interests.	2. *Alternation of Work and School:* Students should alternate time in school and time in a job.	2. *Schools:* High schools should be more diverse, smaller, and use a variety of learning sites.
3. *Careers:* Curriculum should provide work experiences in the community.	3. *Business Alliances:* Business and industry should receive financial support for youth-training programs.	3. *Restructuring:* Governance of schools should encourage broad participation in decisions.
4. *Global Education:* Curriculum should study the whole earth as an interdependent biological and social system.	4. *Youth Communities:* Establish communities where youth would combine education and public service.	
5. *Alternatives:* The school should provide different paths to graduation.	5. *Youth Employment:* Revise legislation to job opportunities where youth will not be restricted, e.g., child labor laws and minimum wage.	

TABLE 1.1 *Summary of Recommendations (continued)*

The Reform of Secondary Education (Brown, 1973)	Youth: Transition to Adulthood (Coleman, 1974)	The Education of Adolescents (Martin, 1974)
6. *Regulations*: There should be more flexibility in governance so alternatives can be introduced, e.g., length of school day and year, daily attendance.	6. *Vouchers*: At age 16, give students a voucher for four years of college education.	
7. *Television*: The influence of TV on students and its potential in the classroom should be investigated.		
8. *Violence*: Schools should have security plans.		
9. *Student Rights*: Schools should eliminate corporal punishment; make rules known to students.		
10. *Compulsory Education*		

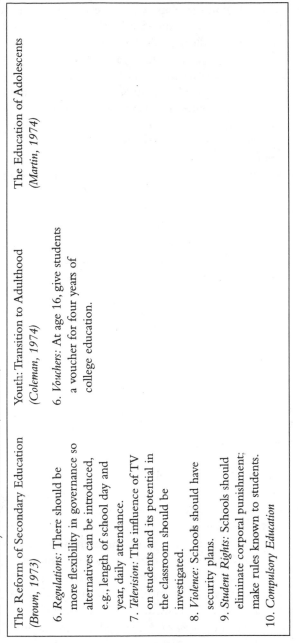

Several curriculum-development projects were initiated in the 1970s. The Biological Sciences Curriculum Study (BSCS) developed the *Human Sciences Program* (HSP) for middle schools, Lawrence Hall of Science published *Outdoor Biological Instructional Strategies* (OBIS), a development team at Florida State University completed *Individualized Science Instructional Strategies* (ISIS), and the *Unified Science and Mathematics for Elementary Schools* (USMES) staff finished their project. These materials represented innovative new programs, in contrast to materials designed to improve extant courses in biology, chemistry, or physics.

Science curriculum developments in the 1970s had only limited success at best for a number of reasons: for example, the unique nature of the programs, reduced NSF support for teacher education, and the lack of national support for educational innovation. Severe criticism was also leveled at the use of public money (usually from NSF) in developing new materials (which were commercially published) and in training teachers to use them. Many objected to the use of public funds in ways that contributed to commercial profit.

The 1970s shift toward conservatism reinvigorated criticism of school programs. *Man: A Course of Study* (MACOS), developed by the Education Development Center (EDC), to introduce middle school children to anthropology, was attacked by conservative groups. When Congress finally held hearings on the project, it voted to eliminate support for the program (Dow, 1991). Work on other programs was scaled back and occasionally stopped.

As a result Congress asked the NSF to examine the status of American science education. What was happening to course improvement programs? What teacher education programs were successful? What was happening in science classrooms? Were science education programs aligned with contemporary science, education, and social demands? In 1976, NSF funded three projects, each with a different perspective, on the status of K–12 science education. At Ohio State University, Stanley Helgeson, Pat Blosser, and Robert Howe reviewed the science education literature for the period 1957–1975, addressing such topics as curriculum, instruction, teacher education, financial support, and needs, trends, and issues (Helgeson, Blosser, and Howe, 1977). At Research Triangle Institute, Iris Weiss (1978) directed a thorough national survey of classroom teachers, science supervisors, and school administrators. The survey elicited information on course offerings, current programs, instructional methods, enrollment, teaching assignments, and support. At the University of Illinois, Robert Stake and Jack Easley (1978) designed and carried out eleven case studies in a variety of schools and communities and then conducted an exhaustive analysis of the case studies. All three projects looked at what educators *recommended,* what they *reported* doing, and what they *really did* in K–12 science education. In addition, the National Assessment of Educational Progress (1978a, 1978b) revealed what students *actually learned.* But even these "snapshots" were a long way from a complete picture of science education.

Norris Harms at the University of Colorado proposed developing an overall picture of the NSF studies and the NAEP reports, an endeavor he called "Project Synthesis" (Harms, 1977). Twenty-three science educators participated in "Project Synthesis," which consisted of five focus groups—biology, physical science, inquiry, elementary school science, and STS—each of which synthesized information from the aforementioned NSF and NAEP reports into a common educational framework consisting of goals, instructional programs, implementation, teacher characteristics, teacher education, facilities and equipment, instructional practices, student characteristics, and evaluating and testing.

The synthesis process used a discrepancy model, which involved developing a desired state of science education for the different curricular domains and common educational features. After identifying the desired state, the teams reviewed the NSF and NAEP reports for the actual state of the same domains and features. The final step in the process was to identify discrepancies between the desired and the actual states. The research teams then made specific recommendations, based on the discrepancies, for reforming science education. These recommendations were presented in a final report to NSF (Harms and Kahl, 1980) and in a National Science Teachers Association (NSTA) report (Harms and Yager, 1981), which received wider distribution and thus had a greater impact. In the 1970s, "Project Synthesis" represented the most significant policy statement in American science education.

The NSTA report included many recommendations. The following list summarizes some of the more important ones.

- Goals must be reformulated along with a new rationale, a new focus, and a new statement of purpose. The new goals must take into account the technological orientation of society and participation by citizens in science-related decisions.

- Curriculum materials must be redesigned to align with the new purposes and goals. The new curricula should include direct student experience, technology, and a focus on personal and social decisions.

- Teacher education (both preservice and inservice) must be revised to support the proposed new directions for science curriculum and teaching.

- Exemplary materials must be identified and made available for science teachers' use.

- Research on science teaching and learning must be translated into programs and practices. The separation of researcher from practitioner cannot continue, and all facets of the profession must work in concert.

- Evaluation must receive renewed attention and be aligned with curriculum and instruction. New approaches to assessment should be incorporated.

- Implementation of new school science programs needs much greater attention and support within local school systems (Harms and Yager, 1981, pp. 129–130).

The report, based on the most thorough analysis of science education carried out during the decade, presented both a reassessment and a comprehensive set of general policy recommendations.

The Third Educational Reform: A Nation at Risk

In 1983, the National Commission on Excellence in Education published *A Nation at Risk*. A weak U.S. economy and strong international businesses, especially in Germany and Japan, signaled a new crisis. Reformers had paid little attention to the transition from youth to adulthood. The nation, it now seemed clear, needed higher levels of academic achievement, a productive workforce, economic progress, and a strong national defense.

A Nation at Risk launched the third wave of postwar educational reform concluding on an ominous note: "The educational foundations of our society are presently being eroded by a rising tide of mediocrity that threatens our very future as a Nation and people." *A Nation at Risk* also put forward remedies to eliminate or at least reduce the risk. Overall, it emphasized "the basics," including three years of science and mathematics, and recommended greater efficiency in use of the school day and a longer school day and school year. The report also recommended higher expectations by colleges and universities and for those entering teacher preparation programs, and gaining support for educational reform from citizens.

During the 1980s, several hundred more reports appeared, each listing school-based problems and agendas for reform. Almost every personal or social issue seemed to have a constituency using that issue to push for reform. In a very real sense, these reports exemplify the popularization of education, the need to think systematically and ecologically, or, in Cremin's (1975) terms, comprehensively, relationally, and publicly. In 1990, President George Bush and the governors of the fifty states adopted six comprehensive goals for education.

- *Goal 1:* By the year 2000, all children in America will start school ready to learn.

- *Goal 2:* By the year 2000, the high school graduation rate will increase to at least 90 percent.

- *Goal 3:* By the year 2000, American students will leave grades 4, 8, and 12 having demonstrated competency in challenging subject matter, including English, mathematics, science, history, and geography; and every school in America will ensure that all students learn to use their minds well, so that they may be prepared for responsible citizenship, further learning, and productive employment in our modern economy.

- *Goal 4:* By the year 2000, U.S. students will be first in the world in mathematics and science achievement.

- *Goal 5:* By the year 2000, every adult American will be literate and will possess the skills necessary to compete in a global economy and to exercise the rights and responsibilities of citizenship.

- *Goal 6:* By the year 2000, every school in America will be free of drugs and violence and will offer a disciplined environment conducive to learning.

Goal 4, on science and mathematics, includes three objectives that should be noted. First, mathematics and science education will be strengthened throughout the system, especially in the early grades. Second, the number of teachers with a substantive background in mathematics and science will increase by 50 percent. Third, the number of U.S. undergraduate and graduate students—especially women and minorities—who complete degrees in mathematics, science, and engineering will increase significantly. Goal 4 has received the greatest publicity, perhaps because it is clearly competitive and appeals to comparisons with other countries. I would argue, however, that Goal 3 presents a much more reasonable, achievable aim, especially since, in defining the content that can be achieved at grades 4, 8, and 12, the *National Science Education Standards* emphasize inquiry and design, abilities that will help students "use their minds well."

In April 1991, President Bush released *America 2000,* a long-term strategy to accomplish these six educational goals. The strategy has four parts, all of which are being pursued simultaneously: first, improving today's schools, making them better and more accountable; second, inventing new schools for a new century; third, encouraging continuing learning for those who have graduated; and fourth, involving communities and families in school programs. *America 2000* sets clear goals and outlines a workable strategy. The policies are vague, but the science education community has an opportunity to make concrete the programs and practices that will achieve these goals. Reform involves not merely a particular program but the entire educational system. The report also calls for what could potentially be broad-based community support for new programs.

America 2000 establishes a deadline for improvement: the year 2000. We need to remind ourselves that the meter is running.

The decade of the 1980s was, in retrospect, an era of reports. Beginning with *A Nation at Risk* in 1983, reports proclaiming the need to improve American education and providing myriad recommendations proliferated. Their number could be seen as a mark of the importance Americans attach to education, but the variety of recommendations was overwhelming and sometimes contradictory. Those associated with the quantity of science education, such as increasing required courses, school days, and the length of the school year, were implemented first because they were easiest. What remains are the more difficult aspects of educational quality and appropriateness: improving and coordinating curriculum, instruction, and assessment, and—especially critical—implementing those changes in the nation's classrooms.

By the late 1980s, there were more than three hundred reports, all admonishing those within the educational system to change. These reports consistently pointed out the specific need for reform in science education (Champagne and Hornig, 1985, 1986, 1987; Champagne, 1988; U.S. Congress, Office of Technology Assessment, 1988a, 1988b; American Association for the Advancement of Science, 1988;

Task Force on Women, Minorities, and the Handicapped in Science and Technology, 1988). Depending on the sponsoring group, recommendations emphasized such issues as updated scientific and technological knowledge, application of learning theory and teaching strategies, different approaches to achieving equity, and better preparation of students for the workplace.

Forces external to science education also caused some educators to rethink school science programs. For example, many of the social concerns of the 1970s were substantiated in the *Global 2000 Report to the President: Entering the Twenty-first Century* (Barney, 1980), which concluded that

> If present trends continue, the world in 2000 will be more crowded, more polluted, less stable ecologically, and more vulnerable to disruption than the world we live in now. Serious stresses involving population, resources, and environment are clearly visible ahead. Despite greater material output, the world's people will be poorer in many ways than they are today. For hundreds of millions of the desperately poor, the outlook for food and other necessities of the life will be no better. For many, it will be worse. Barring revolutionary advances in technology, life for most people on earth will be more precarious in 2000 than it is now—unless the nations of the world act decisively to alter current trends. (p. 1)

Science and technology have moved from a position of benign marginality to one of substantial, if questionable, usefulness. In the 1980s, in addition, the general public began questioning the appropriate balance between science and technology and society. Should there be more government involvement to encourage greater advances and avoid problems or less? This question touches on underlying issues not evident earlier: the greater impact of science and technology on society, for example, and the many social problems that have arisen due to limited resources and increasing global interdependence. Society's perceptions of science and technology have important implications for science education policies, programs, and practices. They highlight a view of scientific literacy that requires more than an understanding of the concepts of traditional scientific disciplines. Citizens need to understand science in a social context, its interdependence with technology, and the nature and processes of both science and technology. These themes (developed more fully in later chapters) set the stage for a "general education" view of scientific literacy and established the perspective of the major policy statements of the 1980s and 1990s, *Science for All Americans* (Rutherford and Ahlgren, 1989), *Benchmarks for Science Literacy* (American Association for the Advancement of Science, 1993), and *National Science Education Standards* (National Research Council, 1996).

The contemporary reform movement in science education is different from that of the 1960s and 1970s in several ways. It is more vigorous, it is fueled by data from national and international assessments, it is more penetrating and pervasive within education, and it is more political. The goal of science education today is scientific literacy for all students. In contrast in the 1960s, the goal was producing more scientists and engineers. Achieving scientific literacy for all students, not simply those training to be scientists and engineers, is, in my view, more closely aligned with the current general education orientation of the schools. But if scientific literacy for *all* students makes contemporary reform more educationally appropriate, it is also more difficult, since it means involving students who are currently underserved, such as the poor, minorities, and women. Producing more scientists and engineers remains an important goal, but it should not be the exclusive orientation of K–12 science programs, which must serve a general education function.

The challenge is to design science programs that accommodate the broad diversity in our schools. This goal confronts all of us in the science education community with the cultural paradox I pointed out earlier: How can we have equality *and* excellence? That is, how can we hope to develop the unique potential of each student and still have common experiences for all students? This issue is one we have not yet resolved, although there has been some thoughtful commentary (Eisenkraft, 1995).

Current reform has focused almost exclusively on policies rather than on programs. It is often equated with publications such as *Benchmarks for Science Literacy* (American Association for the Advancement of Science, 1993) and the *National Science Education Standards* (National Research Council, 1996), which are policy-level statements. The point here is that we have concentrated not on new science programs for schools but on developing new policies. Policies allow more freedom to local, state, or national groups to design science curriculum, but they also involve greater responsibility for incorporating the new purposes set forth in the reports.

The 1990s are a time of widespread educational reform. Unlike the 1960s, when reform began and ended with science and mathematics education, educators today are talking about "school restructuring" and "systemic initiatives." Trying to sustain reform within a status quo educational system was an extraordinarily difficult undertaking. But when the entire educational system is changing, science education becomes an integral participant in that change.

Beginning in 1991, the NSF's Directorate for Education and Human Resources, under the direction of Luther Williams, initiated a series of systemic reform initiatives. This approach established a paradigm vastly more dynamic than earlier science curriculum models. In the 1960s and 1970s, scientists were generally more involved in curriculum projects. Today, although scientists and engineers are involved, business and industry have a greater influence, a situation due, in part, to the fact that issues of national strength, economic productivity, and a literate workforce are driving reform.

Conclusion

Since World War II, the reform in science education has broadened to encompass more components of the educational system. In general, the purposes or goals defined by science educators have determined curriculum programs and teaching practices. In the 1960s, for example, the emphasis was the structure of scientific disciplines and the process of scientific inquiry. In the 1990s, it is scientific literacy for all students, standards, and systemic initiatives. Implementation of the new goals has been uneven. In the 1960s the reform leaders did not formulate policies to link new purposes and new programs. In the 1990s, in contrast, we have policies but few new programs, and for those that have been developed, nothing close to the support and enthusiasm we witnessed in the 1960s.

Postwar science education reveals little evidence of significant or sustained change in classroom practices, although the success of any reform can only be judged by the degree of improvement in science teaching and student achievement. Rethinking science education, restating policies, and redesigning programs are all necessary, but they must not be equated with improvement. The past fifty years have witnessed only limited success in improving science education. Science educators have failed to transform purpose into practice, and they have also consistently underestimated the power of school systems and science teachers to maintain the status quo.

2

Contemporary Reform of Science Education

The last fifty years of science education reveal a pattern of cycles of reform. At least two factors stand out. First, the reforms emphasized *input:* they changed features such as curriculum materials, teaching strategies, and time spent studying science, but they did not identify a clear set of *outcomes* and focus attention on achieving them. I return to this theme in the discussions on *Science Education Standards* (National Research Council, 1996) because this two report shifts the primary emphasis of reform from inputs to outputs, that is, to achieving higher levels of scientific literacy. Second, reform had little effect on teaching and learning in classrooms. We offer a seemingly limitless supply of activities, techniques, and materials, but the cumulative effect is only marginal. We have to ask why this is so.

In the past, the science education community has done an excellent job with the easiest task, that of changing purposes. It has done a good job with the more difficult task of developing instructional materials. But it has done an extremely poor job with the essential task of implementing the new programs and improving instructional practices. We need to see science education as an interdependent system consisting of classroom teachers, school personnel, science supervisors, teacher educators, and other science educators who develop curriculum materials, conduct research, and teach science in colleges and universities. We need to clarify the purposes of science education and then establish continuity among these various components. The overarching aim is a more consistent and coherent system of science education.

A Systems Perspective

A number of discussions of science education seem to suggest a systems view (Bertalanfy, 1968; Laszlo, 1972; Churchman, 1979) and to imply that reform must be approached systemically. For example, the NSF has recommended systemic initiatives for states, urban districts, and rural areas. Although an educational system comprises various components—governance, administration, instruction, and learning experiences (Banathy, 1991)—teachers are the essential component in improving educational achievement. Other school personnel, such as science supervisors, principals, and school boards, as well as nonhuman components like materials, equipment, and facilities are indeed important. Yet the position of science teachers must be central because of the crucial decisions they make about interactions with students, selection of materials, provision of learning opportunities, and the students' subsequent achievement of a higher level of scientific literacy.

A system is a group of interdependent objects that function as a whole (Rapoport, 1968). The four general features of systems are boundaries, components, flow of resources, and feedback (Thorsheim, 1986). One can consider the specific *boundaries* that define a system, for example, systems within the human body. In biology, one can focus on cells or organisms or the biosphere. In science education, one might consider teachers in classrooms, science programs in schools, or science education within one state. Identifying a system's boundaries makes the discussion clear and tacitly recognizes that there are other boundaries that could have been used. Resources for the system must cross these boundaries, and it is at the boundaries that resistance to change is most often encountered in educational systems. A simple example will help. Let us suppose a science department wants to implement a new program. The administration must provide financial support for the program, and the final allocation resources allows the new program to join the departmental subsystem of the school. Many other nonfinancial factors can also influence the adoption of a new program, such as the support within the department, the understanding of other school personnel, the availability of adequate physical facilities and equipment, and political support (or opposition) in the community.

In addition, a systems perspective involves *components* and their effect on the system. Examples of components that might influence the reform of school science programs include the stated and unstated policies of the school and the school district, administrative support, state curriculum frameworks, financial support for state adoptions, staff development initiatives, and teachers' unions.

A third feature of a systems perspective involves the *flow of resources,* which recognizes the place of factors like information, personnel, time, and budgets in the educational system. By examining the flow of resources, especially across boundaries between various components, science educators gain an important perspective. How does information about a new science program, for example, get into a school system? What and where are the barriers that stop or reduce the flow of resources? To understand how to reform science education, science educators should review the interactions that occur at the boundaries between systems (or subsystems) and the flow of resources, such as teaching strategies, curriculum materials, and budgets.

Finally, there is *feedback* within the system. How supportive of reform and innovation is the administration? Do teachers get information about the importance of teaching science and about how well they are doing? Can they depend on budgetary support? Does the school system generally support the teaching of science?

There are two basic types of feedback processes—reinforcing (positive) feedback and balancing (negative) feedback. Reinforcing feedback produces growth and, often, rapid change. Even small changes can accelerate decline when amplified on a larger scale. The decline in bank assets due to a financial panic and the reduction in gasoline supplies during an energy shortage are two examples. In science education, I would argue, the continued response to requests for short-term, single-lesson, how-to activities has a reinforcing effect that works against steady, long-term reform. When there is a goal balancing (negative) feedback operates. If, for example, the goal is to maintain room temperature at 65°F, then balancing feedback will act to adjust the system by turning the heat on or off. The goal can be aligning curriculum materials, making assessments within the curriculum, or developing teaching strategies based on the *National Science Education Standards* or state curriculum frameworks.

One essential outcome of a systems perspective is a change in thinking—recognizing interrelationships rather than cause and effect claims, and process rather than quick results (Senge, 1990a, 1990b). A systems perspective can help lead teachers, department chairs, science supervisors, and others in leadership positions. For too long, we have focused on one component of a system, such as assessment, and assumed that changing that one component would improve the system. But this is unlikely because the system encompasses too many components, a variety of resources, and varied regulatory mechanisms.

The whole business of contemporary educational reform is, in fact, an especially disorganized process. Some states have undertaken reform in assessment, teacher education, curriculum, instruction, school restructuring, and science education simultaneously, but these initiatives often lack continuity and coordination. Science educators at various levels must provide some coherence and coordination within the system, organization, or group they are leading. A systems perspective can provide valuable insights and assistance in these difficult situations.

The Dimensions of Reform

Chapter 1 characterized the 1990s as the decade of reform, and *America 2000* has largely defined the time frame for the reform. Whether we are ready or not, the year 2000 will see a flurry of assessment and pronouncements about how well science educators did in the last decades of the twentieth century. (We can only hope that not every group that published a report during the 1980s and 1990s will feel compelled to produce an assessment for the year 2000.)

Another dimension of reform is school space. In the 1960s, curriculum reform began at the secondary level and then progressed to the elementary level. In the 1990s, reports have generally addressed the whole K–12 span, as has reform at the local level. National funding for new programs has focused on elementary and middle schools first and then high schools. Some policy reports have followed the

TABLE 2.1 *Dimensions of Reform*

	1960s	1990s
Symbol	Man on the moon and space race with USSR	National strength—economics, productivity
Goal	Education and recruitment of scientists and engineers	Scientific and technological literacy for all citizens
Curriculum	National programs primarily supported by NSF	National policies, state frameworks, local development or local implementation of programs
Grade levels	K–12 (physics to elementary, omitted junior high)	K–12 (including middle school)
Instruction	Inquiry and laboratory	Instructional models including laboratory
Evaluation	Standardized tests	Authentic and performance assessment, national and international evaluation
Implementation	NSF institutes for curriculum programs and disciplines	Professional development

same sequence (Bybee et al., 1989, 1990; Champagne and Raizen, 1991). The important point is that school science programs that are structured from the high school level down—from twelfth-grade physics to elementary school science—are very different from science education programs structured according to a more holistic K–12 orientation or one that is unique to a particular level.

The scale of reform, in my view, must borrow a theme from the environmental movement: think globally and act locally. Reform efforts are occurring at the local level, but the NSF has also recommended programs for state, urban, rural, and local systemic initiatives. Many new programs have a national orientation, and beyond these, reform is ongoing in many other countries.

In America today we seem to have a broader perspective than in past educational reform efforts. We are incorporating changes in curriculum, in instruction, and in assessment. Teacher education programs are changing and there is greater recognition of the process of implementing innovative programs and professional development (Hall and Loucks, 1978; Loucks-Hornsby, Stiles, and Hewson, 1996). All these are a significant departure from the curriculum reform of earlier decades. (See Table 2.1 for a summary.)

A Framework for Reform

The following discussion outlines a framework for describing and analyzing reform initiatives. I suggest four basic types of reform initiatives: purpose, policy, programs, and practice. This framework applies to different initiatives, those for example, on a national or local level, within a state or school district, within a college or university, or within a research and development group (for a discussion of this framework in relation to goals, see Bybee, 1979, 1985; Bybee et al., 1992a, 1992b).

Purpose

Purpose includes aims, goals, and rationale. Statements of purpose are universal and abstract, and apply to all concerned with reforming science education. Achieving scientific and technological literacy is an example of a purpose statement.

Policy

Policies are more specific statements of standards, benchmarks, state frameworks, school syllabi, and curriculum designs based on the purpose. Policy statements are concrete translations of the purpose and apply to subsystems such as curriculum, instruction, assessment, disciplines, teacher education, and grade levels within science education. Specification of the knowledge, skills, and attitudes required to improve scientific and technological literacy in all grades is an example of policy. *National Science Education Standards* is such a statement of specifications.

Program

Programs are the actual materials, textbooks, software, and equipment that are based on policies and developed to achieve the stated purpose. Programs are unique to grade levels, disciplines, and types of science education. Curriculum materials for K–6 science and technology and a teacher education program are examples.

Practice

Practice is the specific actions of the science educators. Practice is a unique and fundamental dimension, and it is based on educators' understanding of the purpose, objectives, curriculum, school, students, and their strengths as a teacher.

Reform can begin in different ways. Professional science teachers could change how they understand effective practice and function in the classroom. Professional scientists and science educators could design new science programs that take content, assessment, and instruction into account. Policy makers could set new policies based on what they think citizens need and want. Finally, those who review trends and issues in society and science education could use their analysis to identify new goals and purposes.

The loosely connected levels at which reform operates, then, include purpose, policy, program, and practice. Each level has its own perspective on what is important and how the educational system works. The philosophers who talk of new goals and purposes can argue their case, publish their ideas, and try to persuade others: their influence, however, is limited to persuasion. Policy can set conditions for effective programs and practices, but at the national level, it cannot mandate school science programs and classroom practices. Setting the conditions, however, does influence decisions and can result in mandates at the state and local levels. Setting a curriculum framework and adopting a program within a state or school district directly influences effective practice and provides certain opportunities for students to learn science. It does not control science teachers or science teaching.

Here is the essential issue: we need an educational system that is consistent and coherent, one that has a coordinated set of purposes, policies, programs, and practices. We have to honor the right of states and local school districts to set policies and select programs they think will help their students achieve higher levels of scientific literacy. But state and local school personnel also have a responsibility to design science education systems that achieve common goals like scientific literacy.

PURPOSE Nationally and locally, science educators need a purpose, ideally one that is congruent across the different levels (K–college) and different disciplines. Without a purpose, science education lacks direction, coherence, and coordination. A purpose statement for contemporary reform might be "achieving scientific and technological literacy for all students."

A variation on this aim, albeit less clear and thus weaker, is the outcome articulated at the 1990 Governors Conference: by the year 2000 the United States will be number one in science and mathematics achievement. In this case, the president and the governors have indicated how we will know if we have achieved our purpose (we will be number one in achievement). But this goal does not really clarify what "achievement" consists of or how to develop a base of knowledge, skills, and dispositions. We do have excellent examples of scientific literacy. One of the most thorough definitions is *Science for All Americans* (Rutherford and Ahlgren, 1989), which also develops examples of goal statements for local science programs and state frameworks. Usually the first portion of a syllabus for local school programs is a statement of purpose for science education in that school district or state.

Virtually everyone in science education, at any level, should be able to agree on a purpose such as achieving scientific literacy. Widespread agreement and support provide unity and universal acceptance among science educators in all disciplines and at all educational levels. Paradoxically, the strength of purpose statements— universal agreement—is their abstract quality, which is also their weakness. Those concerned with other, more specific, aspects of science education do not regard purpose statements as helpful or useful. This is why we need to translate purposes into policies.

POLICIES Translating purpose into specifications and requirements introduces policy. The NRC's *National Science Education Standards* (1996) exemplifies national science education policies. Policies can address different aspects of science education, such as curriculum, instruction, assessment, professional development, equity issues, disciplines, grade levels, teacher education, and program implementation; they should represent the broad purpose and the specific concerns, needs, and requirements of the educational component being addressed. They should also inform and regulate the decisions made in the actual development of programs. Policies would, for example, help answer questions about how to achieve scientific and technological literacy through curriculum materials developed for minority students in elementary schools. Although policies are not as abstract as purpose statements, they are not practical for actual classroom use.

Edward Pauly (1992) comments on the limitations of current policies.

> Most of the current school-reform proposals offer little concrete help to teachers and students—the people who make education work in American classrooms. School-choice plans, site-based management, national testing schemes, teacher-certification requirements: they all may have a role to play in responding to the failures of American education, but their proponents are surprisingly silent when confronted with a classroom discipline problem or an alienated, bored class. The school-reform movement has failed to give us policies that provide direct, immediate assistance to the people in our classrooms who do the daily work of teaching and learning. What we need is a different approach to education reform—one that offers policies that are designed with teachers' and students' daily struggles foremost in mind. (p. 36)

Policies provide an essential bridge between purposes and programs, but they lack the usefulness of lesson plans. As Pauly suggests, policies should be designed with more thought to teachers. Science education relies on many frameworks: the *Science Framework for California Public Schools* (California Department of Education, 1990), *Benchmarks for Science Literacy* (American Association for the Advancement of Science, 1993), *New Designs for Elementary School Science and Health* (Biological Sciences Curriculum Study, 1989), reports for the National Center for Improving Science Education (Bybee et al., 1989, 1990), and other frameworks for teaching the history and nature of science and technology education (Bybee et al., 1992a, 1992b) and biology education (Biological Sciences Curriculum Study, 1993). These policy statements are not designed to offer direct help to teachers facing specific classroom problems such as discipline, alienation, or boredom. Pauly has identified a continuing tension between policy makers and classroom practitioners. We need statements about larger purposes that are appropriate for different disciplines (for example, biology), levels (for example, secondary), and issues (for example, equity), and these statements, by their very nature, must have classroom teachers in mind, even if they cannot address every concern of every teacher. The final responsibility for implementing reform lies with science teachers. It cannot be any other way.

Curriculum frameworks, a new feature on the science education landscape, have proven helpful to science supervisors and curriculum developers at local, state, and national levels, but they remain policy statements, a little more concrete than national reports but not as usable as curriculum programs.

Contemporary reform, especially in the 1980s, can be characterized as one of policies rather than programs. Certainly we need policies, but I think we have reached the point of diminishing returns in policy development. Now we need programs, whether they are developed by national groups, such as Biological Sciences Curriculum Study (BSCS), Education Development Center (EDC), Technical Education Centers (TERC), National Science Resources Center (NSRC),

National Science Teachers Association (NSTA), and Lawrence Hall of Science (LHS), or regionally, or locally by school personnel.

PROGRAMS   Science programs should be consistent with policies. They are in fact a concrete representation of policies. The continuity between purpose statements and programs should be strong and well coordinated. Examples of science education programs abound. They include abundant curriculum materials, teacher education programs, and assessment packages.

In the adoption process, many local school districts develop policy statements and then identify programs that best match these policies. When district personnel realize the magnitude of developing their own program, they often find it more reasonable to opt for a program developed by another group and commercially available.

As we saw Chapter 1, one of the major insights of the 1960s reform was that new curriculum materials do not in and of themselves cause a lasting change in science teaching. This is the case even with support through inservice education. The process of implementing an innovative program requires systemic support as well as specific background and strategies for teachers (Hall and Hord, 1987; Hall and Loucks, 1978; Fullan, 1982; Fullan and Stiegelbauer, 1991).

PRACTICES   Any science educator who has made a presentation at a conference has also had the experience of answering teachers' questions about the appropriateness of the purposes, policies, and programs in their specific situation. Edward Pauly's comment, quoted earlier, is an example. The explicit implication is that the purposes, policies, or programs are useless. These questions reinforce the need for the final step in translating purposes to practices. Each classroom is a unique "ecosystem." Each science teacher has particular strengths and skills, understands his or her students (as individuals and as a group), and works within this unique environment. What this suggests is that science teachers have a professional responsibility to adapt materials to their individual classroom situation. Of course, teachers ought to have support, for example, through professional development. That assigns appropriate responsibility to other science educators, such as science supervisors and teacher educators, and to school personnel, such as superintendents and building principals. Adaptations and improvements should also be consistent with local, state, and national policies, especially those programs related to effective teaching whose outcomes inspire popular consensus.

One should assume a systemwide view of this framework. History suggests that the process does in fact proceed in accordance with what is called a "top-down approach," but there is no reason that reform cannot, or should not, originate with professional science teachers as it has with other components in the system. Science education includes philosophers, policy makers, curriculum developers, and classroom teachers, who interact and receive continuous feedback, albeit with varied levels of strength. Coordination problems arise when one group tries to dominate. Each group has common responsibilities that include doing the best job possible at the task of reform and supporting others within the system. All of these points are also applicable to reform at the local level. We need to recognize the distinction between a heavy-handed national to local approach and coordinating the activities and procedures of various districts, schools, and classrooms across the nation. Achieving a national agenda while satisfying local mandates is the challenge of state and local supervisors and science teachers.

Translating the Perspectives

At the interface between these initiatives are science educators, who assume responsibility for translating purposes to policies, policies to programs, and programs to practice. Indeed, their work is absolutely critical to the whole process of science education reform. The essential feature of this process relates to my themes of coordination, consistency, and continuity, and one goal is a coherent educational system. But it must be informed by national perspectives and mediated by local mandates. School personnel working on reform should be familiar with the *National Science Education Standards* and reports such as *Science for All Americans* and demonstrate sensitivity to local issues and concerns.

Individual science educators face a difficult task ensuring consistency and coordination among different components of the system. Each step is like a phase in the transition of matter: it requires energy for activation. By interacting, teacher educators, state and local science supervisors, assessment specialists, and professional developers all facilitate the translation of policies to programs and programs to practices.

Understanding the Difficulties of Reform

National reports provide science educators with the language necessary to justify general policies, broad guidelines, and comprehensive frameworks for curriculum programs and instructional practices. With occasional exceptions, however, the reports seldom translate abstract purposes and policies into concrete programs and practices. If reform is to continue, educators must develop programs consistent with report recommendations, their respective disciplines, and approaches to content (for example, discipline specific or integrated), with the requirements of school systems, and with the requisite needs and interests of students. The development of new programs brings educational reform closer to reality and to the human scale of classrooms.

The aforementioned terms—*purpose, policy, program,* and *practice*—characterize various aspects of educational reform, but any assessment of reform must also examine the changes that result. I equate change with improvement—at the level of science teacher practices and student learning. The real arena of reform is the classroom. It is not the hundreds of national reports, the descriptions of frameworks for curriculum, or the creation and dissemination of new programs supported by the NSF, the Department of Education, Public Health Service, and other public and private agencies. Contemporary reform will only occur when science teachers, science teaching, and student learning change. Yet at the most basic, essential, and important level, that of the classroom, reform is extraordinarily difficult. Why is this so?

Table 2.2 summarizes some of the dimensions of educational reform. If one considers time, scale, space, duration, materials, and agreement on different perspectives, the difficulties become clear. It may take a year or so to develop new statements of purpose and revise goals, but to actually adapt curriculum materials and teaching strategies in response to the new purpose takes much longer, perhaps seven to ten years.

Those who discuss reform often have little understanding of the concept of scale. Occasionally, this even includes scientists and engineers, who should be more knowledgeable. Holding workshops for twenty teachers or visiting a school and talking to students and teachers in order to improve science education are admirable gestures, but the number of school districts (about 16,000), schools (about 110,000), and science teachers (more than 2 million) is too large to assume that such gestures contribute very much unless they consistently focus on one set of messages, such as national, state, or local standards. This does not take into account the critical issue of scale, namely one visit for one hour, compared to the thousands of hours a teacher spends with students over an entire school year.

Let us consider for a moment the space involved in reform, that is, the location of change and the potential impact of that change at that location. Purpose statements abound and have been widely disseminated. The actual purposes of science education—acquiring scientific and technologic knowledge, developing process skills and habits of mind, understanding the personal and social dimensions of science, and examining scientific careers—are common worldwide. The emphasis on various goals and programs, of course, varies considerably. *America 2000* has reached virtually every school, and perhaps every teacher, in the country. *Science for All Americans* (Rutherford and Ahlgren, 1989) is known internationally. Various policy statements, such as the *National Science Education Standards* (National Research Council, 1996) and *Benchmarks for Science Literacy* (American Association for the Advancement of Science, 1993), are also widely known but their impact is concentrated primarily at the national and state levels. This is especially true of legislative policies, which might, for example, mandate additional science requirements for graduation. Programs may extend to the state level because of state adoption lists, but the greatest impact occurs at the level of the local district adopting a new program. Classroom practices manifest themselves at the classroom level.

Once a change has occurred in one of the perspectives, how long does it last? Answering this question provides considerable insight. Goals can change every year. As long as society changes, the goals emphasized will change, and as long as science and technology advance, these goals will undergo continual modification. Why a year? This is about how long it takes to develop, publish, and disseminate an article, report, or book. Establishing policies may take longer, perhaps several years, but they usually last longer, a minimum of five years. State mandates or local syllabi, for example, can change regularly on a cycle of several years. What about the duration of a program once it has been developed and adopted by a school district?

TABLE 2.2 Some Dimensions of Educational Reform

Perspectives	Time	Scale	Space	Duration	Materials	Agreement
	How long it takes for change	Number of individuals involved	Scope and location of the change activity	How long innovation stays once change has occurred	Actual products of the activity	Difficulty reaching agreement among participants
Purpose ■ Reforming goals ■ Establishing priorities for goals ■ Providing justification for goals	*1–2 Years* To publish document	*Hundreds* Philosophers and educators who write about aims and goals of education	*National/Global* Publications and reports are disseminated widely	*Year* New problems, new goals and priorities proposed	*Articles/Reports* Relatively short publications, reports, and articles	*Easy* Small number of reviewers and referees
Policy ■ Establishing design criteria for programs ■ Identifying criteria for instruction ■ Developing frameworks for curriculum and instruction	*3–4 Years* To develop frameworks and legislation	*Thousands* Policy analysts, legislators, supervisors, and reviewers	*National/State* Policies focus on specific areas	*Several Years* Once in place, policies not easily changed	*Book/Monograph* Longer statements of rationale, content, and other aspects of reform	*Difficult* Political negotiations, trade-offs, and revisions

TABLE 2.2 Some Dimensions of Educational Reform (continued)

Perspectives	Time	Scale	Space	Duration	Materials	Agreement
Program	*3–6 Years*	*Tens of Thousands*	*Local/School*	*Decades*	*Books/Courseware*	*Very Difficult*
■ Developing materials or adopting a program ■ Implementing the program	To develop a complete educational program	Developers, field-test teachers, students, textbook publishers, software developers	Adoption committees	Programs, once developed or adopted, for extended periods	Usually several books for students and teachers	Many factions, barriers, requirements
Practices	*7–10 Years*	*Millions*	*Classrooms*	*Several Decades*	*Complete System*	*Extraordinarily Difficult*
■ Changing teaching strategies ■ Adapting materials to unique needs of schools and students	To complete implementation and staff development	School personnel, public	Individual teachers	Individual teaching practices for a professional lifetime	Books plus materials, equipment, and support	Unique needs, practices, and beliefs of individuals, schools, and communities

I would estimate that once a school science program changes, the new version lasts a decade. Once school districts adopt a science program, that is, the textbook and materials remain, in various forms, for at least two cycles of district review and revision. Where adopted, how long did the BSCS, PSSC, CBA, ESCP, SCIS, ESS, and S-APA programs remain in the schools? About a decade, until the schools began considering other textbook programs. Finally, what about the duration of classroom practices? Individual teaching practices probably last several decades, the professional lifetime of most science teachers. Is there a more compelling reason to concentrate on educating the undergraduates who will eventually become the science teachers of tomorrow? Undergraduate science courses provide models of teaching and obviously have tremendous influence.

Materials increase in size and complexity with each perspective. New purpose statements can be presented in articles and reports and other relatively short publications. Policy statements require more expansive formats, perhaps a monograph or a book. Science programs require a complex set of student and teacher materials, including books, courseware, overhead transparencies, and various educational technologies. Classroom practices also require extensive materials as well as equipment and support within the school system. A hands-on, materials-oriented program in elementary schools, for example, presumes some arrangement to supply and replenish kits for teachers.

All these perspectives illuminate the issue of reaching agreement. Developing and publishing new purposes involves a relatively small number of people, and they do not have to agree entirely with the author or authors. Agreeing on more specific policies is more difficult and often requires political negotiations and trade-offs. Adopting a new science program means considering a national agenda, state frameworks, local syllabi, community priorities, budgets, and what is most important, teachers' knowledge, skills, and beliefs about science and science education. But the story does not end there. Once a district agrees on a program, a further difficulty arises, that of accommodating the needs and concerns of the teachers who must implement the program. School personnel must agree on teaching strategies and on how to accommodate the program and the philosophy inherent in earlier agreements.

Even at a general level, Table 2.2 gives the impression that reform has just begun. If we are really serious about improving science education, we should begin thinking about and acting on developing and implementing programs and improving teaching through ongoing professional development.

Table 2.3 illustrates another aspect of the difficulty of educational reform (using technological terms such as *risk, cost, constraints, responsibilities,* and *benefits,* as categories). Moving from the abstract, impersonal scale of the national report to the concrete, personal scale of the classroom, descriptions in the table indicate that vulnerabilities increase dramatically. The responsibilities and requirements of leadership likewise increase. Educators outside the classroom place a tremendous burden on science teachers, often with little recognition of their needs and little support for the tremendous changes required. It is incumbent on every educator who is not in a K–12 science classroom to support those who are ultimately

TABLE 2.3 Some Difficulties of Educational Reform

Perspectives	Risk to Individual School Personnel	Cost to School in Financial Terms	Constraints Against Reform for School	Responsibility of School Personnel for Reform	Benefits to School Personnel and Students
Purpose					
■ Reforming goals	Minimal	Minimal	Minimal	Minimal	Minimal
■ Establishing priorities for goals					
Policy					
■ Establishing design criteria	Moderate	Moderate	Moderate	Moderate	Moderate
■ Identifying criteria for instruction					
■ Developing frameworks for curriculum and instruction					
Program					
■ Developing materials or adopting a program	High	High	High	High	High
■ Implementing the program					
Practices					
■ Changing teaching strategies	Extremely high	Extremely high	Extremely high	Extremely high	Extremely high
■ Adapting materials to unique needs of schools and students					

responsible for reform. Table 2.3 assumes that the new purposes in the many reports on reform must be transformed into policies, programs, and, eventually, practices. The problems increase as these transformations reach science classrooms, the critical level of reform, although, at the same time, the benefits to students also become clear. Science teachers can easily undermine change by demeaning national reports and new science curricula. On the one hand, they resist reform by criticizing others for not being "in the real world"; on the other hand, they do little to change "the real world" that is, according to all reports, outdated and outmoded. The bottom line is low achievement by our students.

Individual perspectives on reform are often constrained by professional responsibilities. For science teachers, these might be the practical requirements of classroom life, and for others, curriculum development and state bureaucracies. We need leadership at all these levels from those in a position to translate and adapt purposes to policies, policies to programs, and programs to practices. Key leaders in this process of adaptation are teacher educators, science supervisors, assessment specialists, professional developers, and school administrators, who can not only function in ways that control and regulate the process of reform and cross boundaries, but also reduce constraints and provide support and feedback for innovative practices. These leaders work to articulate purpose, policy, program, and practice, and their responsibility is to do a thorough and excellent job. They are also responsible for consistency with contemporary goals, such as achieving scientific and technological literacy. Other individuals in leadership positions are responsible for adapting policies to programs and programs to practices.

The term *adapting* implies a change based on a prior structure or on continuity between, for example, policies and programs. Some individuals or groups seek to match another set of policies with extant programs. In such cases, the link between the respective policies and programs must be evaluated in terms of overall purpose.

Reform in Science Education

The frameworks described here provide a useful assessment of progress in transforming science education, although one can also get a sense of progress by determining the time, budget, and effort being spent on the various initiatives. Is the school district spending more time, more money, and more effort on developing its goal statement or on implementing the new science program? At the national level, is funding concentrated on policy statements or on various curriculum materials and professional development programs? I have observed considerable policy effort at the national and state levels. At the local level, science committees often spend little time discussing and debating philosophy and goals; more time on policies for elementary, middle, and high school; a great deal of time developing curriculum materials or deciding which program to adopt. But they devote very little time, money, and effort to implementing a science program or to professional development once a program has been developed or adopted. (I would be happy if many agencies and school districts prove this analysis wrong.)

Elsewhere I have proposed a general process of reform and transformation in science education (Bybee, 1977, 1982, 1994). Here, I want to update those general ideas and place them in a more contemporary context, particularly at the local level. Yet it may not be popular discontent that initiates local reform but the cycle of school district review and revision of programs.

POPULAR DISCONTENT Change in school science education occurs when a critical mass of individuals becomes discontented with current purposes, policies, programs, and practices. Their discontent originates in specific issues such as achievement test scores, national economic decline, national defense, and disparities between advances in science and technology and current programs. A decade of reports on problems in education, including science education, suggests a degree of popular discontent.

NEW DIRECTIONS Awareness of the need for reform initiates questions about the direction, purpose, and focus of the reform. Today, the prominent organizing theme for reform is scientific literacy for all students. *Science for All Americans* (Rutherford and Ahlgren, 1989) clarifies the new emphasis. I would also argue that this new direction folds science education into the larger context of general education (Boyer and Levine, 1981; Hlebowith and Hudson, 1991).

NEW POLICIES Once they have established a direction, science educators must begin to translate the general ideas—for example, the themes and topics from *Science for All Americans* (Rutherford and Ahlgren, 1989)—into specific plans for different grade levels, different institutions, and different disciplines. *Standards* (National Research Council, 1996), *Benchmarks* (American Association for the Advancement of Science, 1993), or indicators also have a vital role as new ideas are embodied in programs.

NEW PROGRAMS The next step involves developing curriculum materials and science programs. The unique circumstances of states, regions, and school districts require a diversity of materials and programs. A great failing of some developers of curriculum materials and textbooks is their view of the program as the solution, whatever unique situation exists in states and school districts. Let me use my own work as a curriculum developer as an example. BSCS programs, such as the elementary program *Science for Life and Living,* the middle school program *Middle School Science and Technology,* and the high school program *Biological Science: A Human Approach* are all excellent programs, but they are not appropriate for all school districts and all science teachers. At the same time, even in those districts where there is congruity between the new policies and the proposed science program, it is also essential to adapt the program to the unique needs of teaching personnel and students.

PROGRAM IMPLEMENTATION The last and most essential step in reform is that of implementing new programs. It may sound simple, but as my analysis indicates, this is the most complex and difficult step. To be sure, reforming science education does not adhere to a neat five-step process. Many changes are occurring at different levels simultaneously. However, we can take a larger, holistic perspective and ask where most activity is occurring and what it is that we need to focus on now. (What we do *not* need are any more reports from groups discontented with the present state of science education.)

What about new directions? *Science for All Americans* (Rutherford and Ahlgren, 1989) provides a major organizing statement, although some elaboration of themes and topics may be needed as well as further clarification of the place of science education in general education. The concept of scientific literacy also calls for further clarification. Nevertheless, this is mere fine tuning. (Some of the aforementioned issues are addressed in later chapters.) Do we need new policies? Several groups and many states have prepared policy statements and curriculum frameworks for science education. The *National Science Education Standards,* the dominant policy statement, should initiate responses across the country and help coordinate and bring continuity to various aspects of reform; and, eventually greater coherence to the educational system. So, do we need new programs? In a very real sense we already have many science programs that meet the reform criteria. We still need new materials, however, especially in light of national standards.

Table 2.4 presents summaries of ten categories common to science education programs over time. It is based on the original work of Paul DeHart Hurd in *New Directions in Teaching Secondary School Science* (1969) later modified by Chris Dede and Jay Hardin (1973). (An earlier version of this chart appears in Bybee, 1992.)

What about implementation? This aspect of science education reform has only just emerged, and it is absolutely vital.

People and Power

Nobody ever said reforming science education would be easy. As the discussion thus far suggests, the whole business of educational reform becomes much more difficult and complex as it moves from abstract statements to actual classroom practices. Underlying this point is a simple truth. Reform requires people to change. The fundamental arena of reform is the science classroom, where change will be evident in the curriculum materials and instructional strategies of teachers and in the learning outcomes of students. Hundreds of additional reports will make little difference. New policies will help some, and new programs will help quite a bit more. But ultimately, it is improved science teaching and student learning that will make the difference.

What does it take to improve science teaching? In the 1960s and 1970s, the United States expended billions of dollars on science teacher education with one relatively simple goal: to improve science teaching through a new inquiry-based curriculum. NSF curriculum materials and university scientists supported this goal. Did we achieve it? By most reports, we did not (Weiss, 1978, 1987; Costenson and Lawson, 1986; Mullis and Jenkins, 1988; Welch et al., 1981).

Let me give another example. In 1988, I reviewed the role of science textbooks

TABLE 2.4 *Reforming Science Programs: Past, Present, and Future*

Past to 1950s	*1960s to 1980s*	*1980s to the Year 2000*
1. Personal/social goals were somewhat recognized.	1. Personal/social goals were largely unrecognized.	1. Personal/social goals are recognized as part of scientific literacy.
2. Scientific knowledge was presented in a logical progression.	2. Scientific knowledge was presented as the "structure of the discipline."	2. Scientific knowledge will be presented in the contexts that have meaning for students.
3. The "scientific method" was presented as a specific procedure.	3. Scientific methods were presented as inquiry processes designed to involve students in "pure science."	3. Scientific and technological methods will be presented as inquiry and design into questions and problems, respectively.
4. Laboratory work was intended to demonstrate, visualize, or confirm knowledge.	4. Laboratory exercises were designed to develop inquiry skills and to "discover" knowledge (mostly reductive analysis).	4. Laboratory exercises will provide opportunities to answer questions, solve problems, and develop decision-making skills.
5. Science programs were determined primarily by textbook and authors.	5. Science programs were determined by curriculum developers and scientists.	5. Science programs will be based on standards and determined by teachers with assistance from curriculum developers, supervisors, and national organizations.
6. The textbook was the curriculum.	6. The textbook and the laboratory were the curriculum.	6. The textbook, laboratory, simulation games, community experiences, electronic media, and other informal educational resources will be the curriculum.

TABLE 2.4 *Reforming Science Programs: Past, Present, and Future (continued)*

Past to 1950s	*1960s to 1980s*	*1980s to the Year 2000*
7. Science was related to technology much of the time.	7. The science-technology relationship was largely neglected for "pure" science.	7. The interdependence of science, technology, and society will be included.
8. Science was presented as established knowledge.	8. Science was presented as an ever-changing body of knowledge that is updated through inquiry processes.	8. Science will be presented as an ever-changing body of knowledge having important influences on society. Updating and using the knowledge for democratic participation will be underscored.
9. A disciplinary and multidisciplinary (within scientific disciplines) approach was used.	9. A disciplinary approach is used exclusively.	9. A disciplinary and interdisciplinary approach will be used at different grade levels.
10. Careers were represented by stereotyped male scientists in the laboratory.	10. Career information in science was largely ignored; programs were primarily directed toward science and engineering.	10. Career information will be directed to multiple scientific and technological occupations for many citizens.

for a National Academy of Sciences publication on biology teaching (Bybee, 1989). My findings yielded no surprises: many individuals are quite critical of current textbooks. What interested me was the discovery that science teachers find little wrong with their textbooks and are, in fact, satisfied with them (Weiss, 1987). In trying to explain these observations, I came to the conclusion that it is not new materials or national reports that make the difference for most science teachers. Something much deeper and more important is influencing the direction of science teaching. Resistance to change must be related to other factors. One could be that science teachers do not know or do not share the larger purposes of science education stated in so many reports. It could also be that the needs of science teachers are largely met by current practices, or that the school system does not *really* support any sustained effort to improve science teaching. There is probably an element of fact in all three. In general, science teachers do not review reports and have long forgotten the history and philosophy of education. Personal needs, such as self esteem, of science teachers can be fulfilled through lecturing, and thus demonstrating a knowledge of science, rather than through effective teaching and enhanced student learning. Finally, school systems do not generally support sustained, long-term professional development for science teachers or improvement of science education.

I also am sympathetic to the fact that many science teachers are confronted with students who do not want to learn, who do not have basic skills, and who are often disruptive. Harold Hodgkinson (1991) made this point in a penetrating essay. In this essay, Hodgkinson uses a *leaky roof* metaphor: although we have tried to repair our educational house, we cannot attend to other problems, many of which do not have their origins in schools. Hodgkinson cites an example: "The fact is that at least one-third of the nation's children are at risk of school failure even before they enter kindergarten. The schools did not cause these deficits, and neither did the youngsters" (p. 10).

He goes on to cite a common litany of problems—poverty, addiction, single parenthood, lack of adult supervision, poor medical care, homelessness, abuse, and neglect. Hodgkinson concludes, "I hope readers take away from this article two main points. First, . . . American children are truly an 'endangered species.' And second, educators alone cannot 'fix' the problems of education, because dealing with the root cause of poverty must involve health care, housing, transportation, job training, and social welfare bureaucracies" (p. 16). We cannot, and should not, expect teachers to accomplish this reform on their own. They need the support of virtually everybody associated with the educational system.

In the last thirty years, science teachers have gained significant power. In many respects, they are now, to use a contemporary term, "empowered." Power is the capacity to ensure the outcomes one wishes or expects and to avoid those that one does not wish or expect (Gardner, 1990). Society has empowered science teachers to guide students in achieving scientific literacy. The teacher-proof programs of the 1960s were a misguided attempt to decrease the power of teachers and were rightfully rejected. Today, however, because of unions and their negotiating strength, teachers have more power. Curriculum developers are no longer trying to design programs that circumvent the teacher's role as decision maker. At the same time, the change in power relationships has consequences: science teachers are responsible for what happens in classrooms and accountable for what students learn.

Science teachers should realize that every time they implement a new science program, improve instruction, and make science education meaningful for their students, they are exercising their power appropriately. Every time they are unaware of new goals and programs, rely on how-to-do-it lessons, and are not implementing effective teaching strategies, they are misusing or giving up their hard-earned power.

In *The Predictable Failure of Education Reform,* Seymour Sarason (1991) confronts the whole issue of the power structure in school systems. He points out that we have to view the issue of education reform as a systemic change. We cannot simply *tinker* with components of the system—the length of the school day or textbooks—and expect to achieve any significant reform. We have to recognize the power of relationships within the system and distribute power to those who can make changes that count.

Conclusion

Reforming science education presents every science educator with challenges. The task is not easy and there are no simple solutions. We have completed work on the easiest dimensions of reform—identifying new purposes and formulating new policies. The more difficult tasks, developing and implementing programs, are largely ahead as we approach the year 2000.

3

Searching for Scientific Literacy

"Searching for scientific literacy" suggests an ambiguous process that might, or might not, be concluded successfully. In the United States, this search has continued for more than two hundred years and will no doubt extend far into the future. The search originated with the founding of America, when unnamed and unremembered colonial teachers first asked, "What should we teach our students about natural philosophy?" and "Why should we teach natural philosophy?" The endless interaction between science education and society requires us, from time to time, to ask and try to answer what I have termed the Sisyphean question: What should the scientifically literate person know, value, and do—as a citizen? Attempting to answer this question keeps the cycle of reform in motion (Bybee, 1979, 1994; Bybee and DeBoer, 1993; DeBoer and Bybee, 1995).

The idea of scientific literacy has been a key factor in the formation of school science programs throughout our history. Science educators must establish a working definition appropriate to contemporary reform. This chapter examines the theme of scientific literacy as presented by science educators since the 1950s and its development from a slogan to an elaborate definition, in the *National Science Education Standards* (National Research Council, 1996). The term has been increasingly used as a shorthand version of the fundamental goal of science education. It expresses a set of aims that include traditional subjects (such as physical, life, and earth science concepts) plus content (such as nature and the history of science, the nature of technology, and science-related social issues).

Perspectives from Science Education

A statement such as "achieving scientific literacy for all learners" expresses a general purpose that is persuasive and even patriotic. Who could be against such a goal? On the other hand, exactly what does "scientific literacy" mean? What does the phrase "for all learners" mean? The number of definitions of scientific literacy in the literature requires some review and synthesis. This chapter does not review the entire literature on scientific literacy, but rather, attempts to describe those positions and reports that contributed substantively to a definition and conceptual framework applicable to contemporary curriculum and instruction.

In papers examining historical and philosophical insights about scientific literacy, Paul DeHart Hurd (1987, 1990), who has been writing on the subject for nearly four decades, points out that historically, individuals such as Francis Bacon in the seventeenth century and Benjamin Franklin and Thomas Jefferson in the eighteenth have claimed that what we now call scientific literacy is an educational goal. According to Hurd, the meaning for scientific literacy can be found in the ethos of modern science, in the achievements of science and technology, in social history, in the foundations of education, and in the nature of contemporary society.

The 1950s: The Birth of a Slogan

For decades, individuals have written about what those *not* bound for scientific careers need in the way of science education. In 1946 James Bryant Conant published an article entitled "The Scientific Education of the Layman," in which he addressed the general topic of scientific literacy and made the case for "understanding science," an aim much broader and deeper than the accumulation of scientific information, and focused on the "tactics and strategy of science." Although this article set the stage for Conant's historical case study approach and for later discussions of general education in a democratic society, he did not, at this early date, use the term *scientific literacy.*

According to my research, Conant first used the term in 1952 in *General Education in Science,* a volume edited by I. B. Cohen and Fletcher Watson. In the foreword Conant discusses the need for individual citizens to appraise experts and their advice. "Such a person might be called an expert on judging experts. Within the field of his experience, he would understand the modern world; in short, he would be well educated in applied science though his factual knowledge of mechanical, electrical, or chemical engineering might be relatively slight. He would be able to communicate intelligently with men who were advancing science and applying it, at least within certain boundaries. The wider his experience, the greater would be his scientific literacy" (p. xiii).

In the early 1950s, the term *scientific literacy* was often associated with discussions of general education in science (see Cohen and Watson, 1952). Conant used it to express a broad understanding of science, but to the degree that he defined it, that understanding was contextual. He did not further elaborate on its meaning.

It was Paul DeHart Hurd who first used the term as a major theme for science education. In a 1958 article entitled "Science Literacy: Its Meaning for American Schools," Hurd explained scientific literacy as an understanding of science and its applications to social experience. Science had such a prominent role in society, he argued, that economic, political, and personal issues could not be considered without reference to it. It is worth noting that Hurd's article appeared shortly after the 1957 launch of *Sputnik* and in the earliest phases of curriculum reform. In the first sections of this article, Hurd makes the connection between science and society and then describes the emerging reforms, outlining many attributes of curriculum in what might be called an informal definition. Because of the historical significance of this article, I quote Hurd's comments at some length.

> There is the problem of building into the science curriculum some depth and quality of understanding. It is essential to select learning materials that are the most fertile in providing opportunities for using methods of science. Further efforts are required to choose learning experiences that have a particular value for development of an appreciation of science as an intellectual achievement, as a procedure for exploration and discovery, and which illustrate the spirit of scientific endeavor. (pp. 14–15)

Hurd would also include the history of science if "it should be presented in its more significant aspect as a major intellectual accomplishment of mankind" (p. 15).

Finally, Hurd draws the connection between science and society.

> Today most aspects of human welfare and social progress are in some manner influenced by scientific and technological innovations. In turn, scientific knowledge establishes new perspectives for reflection upon social problems. The ramifications of science are such that they can no longer be considered apart from the humanities and the social studies. Modern education has the task of developing an approach to the problems of mankind that considers science, the humanities, and the social studies in a manner so that each discipline will complement the other. (p. 15)

Although others had employed the term *scientific literacy* before 1958, I believe that Hurd enhanced its use among science educators in the 1960s and formally introduced the term in contemporary usage.

The 1960s: Exploring the Idea

In 1960, Frederick Fitzpatrick edited a small volume entitled *Policies for Science Education,* part of the Science Manpower Project instituted in 1956 at Teachers College, Columbia University. Several contributors used the term *scientific literacy* in their essays, and as Douglas Roberts (1983) points out, they used it as a rallying symbol for a major educational aim. In the first chapter, Fitzpatrick counters the discussion of the need for scientists and technologists and the role of science education with his own argument that science education is for all citizens: "In considering the need for scientific manpower, however, we should not lose sight of the fact that no citizen, whether or not he is engaged in scientific endeavors, can be literate in the modern sense until he understands and appreciates science and its work. . . . If the zeitgeist is to be favorable to the scientific enterprise, including both academic and industrial programs, the public must possess some degree of scientific literacy, at least enough to appreciate the general nature of scientific endeavor and its potential contributions to a better way of life" (1960, p. 6).

As we will see, in the search for scientific literacy the path to sloganeering and the path to a more careful definition may have diverged as early as 1960. Using slogans in education is important and justifiable, but a counterbalance of acceptable and useable definitions is also necessary; otherwise terms assume whatever definition an individual wishes and eventually lose their meaning and their power.

By the early 1960s, other scientists had begun to employ *scientific literacy* to describe the purposes of science education. The physicist Polykarp Kusch published a paper in 1960 entitled "Educating for Scientific Literacy in Physics," which he had presented the previous year at a conference on "Reconstructing Scientific Education for All." He begins by stating that every person should attempt to understand science.

> The attempt, honestly undertaken, almost certainly will lead to scientific literacy if not to profound knowledge. It may lead to a high respect for the methods, the integrity, the spirit, and the results of science. That citizen who respects the structure of science, who is able to view the results of science as a critical and careful statement of man's best knowledge of the behavior of nature is, to my mind, better able to participate effectively in the conduct of our national and international affairs—indeed in every aspect of our life. (p. 199)

Kusch calls for a larger view of the "spirit and nature" of science and draws a connection between scientific knowledge and good citizenship. Other scientists and science educators (Evans, 1962) explored similar themes.

Philip G. Johnson's (1962) article, "The Goals of Science Education," likewise claimed a need for scientific literacy. It must be based, Johnson argued, on scientific knowledge, but that knowledge is much broader than a simple mastery of facts. Johnson added a dimension, that of attitudes and values, particularly those "habits of mind" that extend from the nature of science itself. In reviewing the then-current state of science education, he pointed out the fact that many national committees and individuals were expressing divergent goals, all under the theme of scientific literacy, and suggested that scientists might be stressing a goal, the processes of science, that was "relatively esoteric to science teachers and curriculum leaders" (p. 243). In commenting on the curriculum reform of the 1960s, Johnson concluded that "some goals of science education have become so dominant that the pursuit of other important goals has been severely inhibited: often there has been a failure to recognize adequately the abilities and needs of the general citizen, and thereby to skirt the goal of scientific literacy" (p. 244). Johnson's point also applies to contemporary reform. In the 1960s we skewed the goals of science education—and subsequently our programs—to such a degree that the more general aim, science education for all students, was not achieved. Contemporary reform should avoid repeating the error.

In 1963, Alma Wittlin presented a position on scientific literacy in the elementary school in the journal *Science Education*. First she outlines the requirements of scientific literacy:

- a variety of information

- general and broad rather than deep knowledge

- some understanding of interrelations between fields of knowledge

- a realization of the contribution made by science to human welfare

- appreciation of the bold intellectual adventure implied in scientific discovery

and then argues that individuals need to understand two categories of phenomena, matters of nature and man-made things—technology—because these two make up the environment. In developing a rationale for initiating scientific literacy in the elementary grades, she turns to developmental psychology. In my view, Wittlin's essay is one of the most comprehensive early discussions connecting scientific literacy, developmental psychology, and curriculum development in science.

Also in 1963, Robert Carlton of the NSTA surveyed scientists and science educators on how they defined the term *scientific literacy* and how they thought scientific literacy could be improved. Only a few respondents identified the science and society theme; most recognized and recommended greater content knowledge in specific disciplines and some understanding of scientific methods and accomplishments. Paul DeHart Hurd's definition is worth noting: "A scientifically literate person has a precise understanding of some of the key concepts, laws, and theories of science. He can relate these concepts, laws, and theories in a logical and coherent manner and can appreciate their significance" (Carlton, 1963, p. 34).

Hurd also touches on the personal and social relevance of science. "To be scientifically literate is to understand the place of the individual in the process of discovery and to recognize how the temper of the times influences the evolution of ideas. The scientifically literate recognize the limitations of science and know about its many unresolved problems. Most of all, they appreciate how the use of intelligence in inquiry and experimentation has advanced man's understanding and influenced the course of society" (p. 34).

Charles Koelsche (1965) pursued another conception of scientific literacy, namely, that it comprised the knowledge and skills required to read and understand science as it is presented in the media. Koelsche's study identified 175 scientific principles and 693 vocabulary words in a sample of magazines and newspapers. Koelsche recommended that science programs include these principles and this vocabulary, because together they constituted scientific literacy. Along with Ashley Morgan, Charles Koelsche later published *Scientific Literacy in the Sixties* (1964), which took a functional approach similar to that espoused by E. D. Hirsh (1987) in *Cultural Literacy*.

Many prominent science educators and NSTA began using the term *scientific literacy* during the 1960s as an expression of overall purpose. In 1963, Morris Shamos, a noted science educator, characterized scientific literacy as knowing science in a humanistic way, in other words, feeling comfortable reading or talking with others about science on a nontechnical level. His article previewed later publications (Shamos, 1984, 1988a, 1988b, 1990, 1995) in which he asserted that scientific literacy is essentially unachievable. But here, he took the position that an educated individual should know science in a humanistic way, which includes knowing the difference between science and technology, and understanding the major conceptual schemes of science—the kinetic theory of heat, the molecular theory of matter, the principles of conservation, the gene theory of heredity, and the nature of science.

In *Theory into Action,* an influential document published in 1964, the Curriculum Committee of NSTA suggested that "Science teaching must result in scientifically literate citizens" (p. 8). By "scientifically literate" they meant a "person [who] knows something of the role of science in society and appreciates the cultural conditions under which science survives, and knows the conceptual inventions and investigative procedures" (p. 9). *Theory into Action* presented a strong case for using the conceptual schemes of science and the processes of science as the primary basis for curriculum planning.

The Mid-1960s: An Initial Synthesis

By the mid- to late-1960s, the term *scientific literacy* as clarified by various individuals and organizations, expressed the goals of science education (Gatewood, 1968; Hurd and Gallagher, 1966; National Science Teachers Association, 1964). Yet what of the national reform effort that was in progress? Some reformers (for example, Karplus, 1964) had addressed the general issue of scientific literacy, but most had not, or, if they had addressed the issue, they designed curriculum materials with a different purpose. By the mid-1960s, critics began pointing this out.

In a literature survey of the theme of scientific literacy, Milton Pella, George O'Hearn, and Calvin Gale (1966) examined one hundred articles to determine what science educators meant by the term. They reported the six most frequent referents (the numbers after the item indicate how often the item was used):

1. interrelations between science and society (67)

2. ethics of science (58)

3. nature of science (51)

4. conceptual knowledge (26)

5. science and technology (21)

6. science in the humanities (21)

They found that the primary purposes of this goal were to prepare scientists, to provide the background for careers in technical occupations, and to provide the general education science background required for effective citizenship. Pella, O'Hearn, and Gale used a pyramid model to represent the proportion of the general population served by each of the three purposes. The smallest, near the top of the pyramid, represented scientists. The middle segment represented technological careers. The largest, the base, represented general education in science for effective citizenship. The authors concluded that the greatest emphasis in science education should be on the lowest portion of the pyramid, general education and general population.

Pella used the results of his study to propose a definition of scientific literacy.

1. The scientifically literate person should understand the interrelationships between science and society.

2. The scientifically literate person should understand the methods and processes of science.

3. The scientifically literate person should have a knowledge of fundamental science concepts or conceptual schemes.

4. The scientifically literate person should understand the difference between science and technology.

5. The scientifically literate person understands the relationships between science and the humanities or, better still, looks upon science as one of the humanities.

Pella then assessed the extent to which the new NSF-supported science curricula included the issue of scientific literacy or topics related to it. In general, Pella deemed the new programs inadequate because they included little or nothing on science and society, technology, and science in the humanities (Pella, 1967). To be fair, programs that were developed after Pella's initial review did incorporate these themes—Rutherford, Holton, and Watson's (1970) *Project Physics,* and *The Man-Made World* (Engineering Concepts Curriculum Project, 1971). By the end of the decade, there was a significant disparity between extant science programs and the goal of scientific literacy. Table 3.1 illustrates some of the characteristics of scientific literacy as they were defined in the 1960s.

The Early 1970s: National Recognition

In January 1970, Paul DeHart Hurd published another major article on scientific literacy, although this time he used the term *scientific enlightenment*.[1] As I pointed out in Chapter 1, in the late 1960s and early 1970s, alienated youth received considerable attention. As Hurd commented, "We are only beginning to identify the components of a scientific literacy and those of a human literacy to provide a wholeness to the study of science in schools. This is the relevancy the younger

---

1. In personal correspondence, Paul DeHart Hurd explained that he used *scientific enlightenment* because it reflected the spirit of the times. He said that this article represented a refinement of the 1958 statement.

TABLE 3.1 Some Characteristics of Scientific Literacy: The 1960s

National Science Teachers Association (1964)	Paul Hurd and James Gallagher (1966)	Milton Pella (1967)	National Science Teachers Association (1964)
Conceptual Schemes			Processes of Science
1. All matter is composed of units called "fundamental particles." Under certain conditions, these particles can be transformed into energy and vice versa.	1. Appreciate the socio/historical development of science.	1. Interrelationships between science and society.	1. Science proceeds on the assumption, based on centuries old experience, that the universe is not capricious.
2. Matter exists in the form of units that can be classified into hierarchies of organizational levels.	2. Aware of the ethos of modern science.	2. Ethics of science.	2. Scientific knowledge is based on observation of samples of matter that are accessible to public investigation in contrast to purely private inspection.
3. The behavior of matter in the universe can be described on a statistical basis.	3. Understand and appreciate the social and cultural relationships of science.	3. Nature of science.	3. Science proceeds in a piecemeal manner, even though it also aims at achieving a systematic and comprehensive understanding of various sectors or aspects of nature.
4. Units of matter interact. The basis of all ordinary interactions are electromagnetic, gravitational, and nuclear forces.	4. Recognize the social responsibility of science.	4. Conceptual knowledge.	4. Science is not, and will probably never be, a finished enterprise, and there remains very much more to be discovered about how things in the universe behave and how they are interrelated.

TABLE 3.1 *Some Characteristics of Scientific Literacy: The 1960s (continued)*

National Science Teachers Association (1964)	*Paul Hurd and James Gallagher (1966)*	*Milton Pella (1967)*	*National Science Teachers Association (1964)*
Conceptual Schemes			Processes of Science
5. All interacting units of matter tend toward equilibrium states in which the energy content (enthalpy) is minimum and the energy distribution (entropy) is most random. In the process of attaining equilibrium, energy transformations or matter transformations or matter-energy transformations occur; nevertheless, the sum of energy and matter in the universe remains constant.		5. Science and technology.	5. Measurement is an important feature of most branches of modern science because the formulation and establishment of laws are facilitated through the development of quantitative distinctions.
6. One of the forms of energy is the motion of units of matter. Such motion is responsible for heat and temperature and for the states of matter: solid, liquid, and gaseous.		6. Science in the humanities.	
7. All matter exists in time and space, and since interactions occur among its units, matter is subject in some degree to changes with time. Such changes may occur at various rates and in various patterns.			

generation so urgently seeks." He begins with the curriculum: "The curriculum has not considered any direct way the relation of science to the affairs of man, the actualities of life, and the human condition. The scientific enterprise as a part of general education has meaning in a cultural and social context" (p. 95). Hurd makes a clear connection between the science curriculum and the goals of general education. He also points out that the significance of science concepts in general education is greater for problems of living than for basic research. However, the approach he recommends does not neglect basic concepts, laws, theories, or methods of science, as his essay in Carlton's (1963) survey of scientific literacy indicates.

In 1970, Donald Daugs also published a short article in which he proposed that failure to achieve scientific literacy was due largely to the lack of a definition. He cited Hurd (1963, 1970), along with Pella, O'Hearn, and Gale (1966), as the basis for a workable definition, but what is more important, he suggests that scientific literacy is not an either/or situation but a matter of degrees of achievement. This insight—expanding the definition of scientific literacy—was crucial in later discussions, from various perspectives, of a definition.

Thomas Evans (1970) also considered the characteristics of the scientifically literate person, modifying and elaborating on Pella, O'Hearn, and Gale (1966). His main theme, however, was science teachers' responsibility for developing scientific literacy, and he acknowledged that the "persons responsible for organized instruction in science have not acquired a mature scientific literacy of their own" (p. 82).

The national prominence of the whole issue of scientific literacy increased among science educators in 1971, when NSTA published the position statement "School Science Education for the 1970s." The position statement declared that "The major goal of science education is to develop scientifically literate and personally concerned individuals with a high competence for rational thought and action" (p. 47) and included a list of characteristics. The scientifically literate person, for example, uses science concepts, processes, skills, and values in making everyday decisions as he interacts with other people and with his environment, and understands the interrelationships among science, technology, and other facets of society, including social and economic development (pp. 47–48).

The 1971 NSTA position statement returned to the theme of scientific literacy in different educational contexts. It declared, for example, that "to promote scientific literacy, science curricula must contain a balanced consideration among conceptual schemes, science concepts, and science processes including rational thought processes, the social aspects of science and technology, and values deriving from science." And again: "The achievement of scientific literacy should be the basis for setting objectives, selecting content, learning experiences, and methodology, and for developing a system of evaluation." The NSTA position was indeed quite broad, including history and social issues.

A 1972 publication for the National Association of Secondary School Principals, under the alarming title "Scientific Literacy: Imperative for Survival," contained articles on different aspects of the curriculum (Sonneborn, 1972; Patton, 1972; O'Hearn, 1972) and other fundamental issues, such as science and general education (McQuigg, 1972) and science education in developing countries

(Kando, 1972). The title overstates the case, but the essays present a good overview and direct attention to a consideration of the nature of curriculum and instruction.

The Mid-1970s: Elaborating the Idea

By the mid-1970s, those involved in science education consistently referred to scientific literacy and the issues and challenges associated with it (Smith, 1974). By this time also, the term had assumed the characteristics of a slogan, although some extended the work of Milton Pella and his colleagues in an attempt to further elaborate on the concept.

In 1974, Michael Agin published a conceptual framework for scientific literacy based on a review of the literature. He proposed six broad categories—science and society, the ethics of science, the nature of science, the concepts of science, science and technology, and science and the humanities—to aid in planning interdisciplinary teaching units, describing each and providing examples of topics and units. In his view, "Individuals should become aware of the products and processes of the scientific enterprise within a social setting rather than a social vacuum. Initially, they may view science as being composed of unrelated domains of 'product,' 'process,' and 'society.' But as they become more mature, they should become increasingly aware of the interrelatedness of these domains. Finally, the scientifically mature individual should view science as a social activity with interrelated and interdependent concepts, methods, applications, and influences—a truly scientifically literate person" (p. 414).

Agin's framework is significant for several reasons. First, it is based on a review of the extant literature and thus incorporates the ideas of science educators and scientists, as well as professional organizations. Second, Agin provided good examples of the concepts and of ways to plan and teach science.

Michael Agin organized a symposium on scientific literacy in 1975, for the annual meeting of the National Association of Research in Science Teaching (NARST). Milton Pella (1976) made a presentation, as did George O'Hearn (1976). Pella and O'Hearn worked together at the Scientific Literacy Center, University of Wisconsin, Madison for over a decade, contemplating and carrying out research on the complex issue of scientific literacy (Klopfer, 1976).

George O'Hearn (1976) offered an operational definition of scientific literacy that can be summarized as (1) basic scientific knowledge, (2) the nature of science, (3) the processes of science, and (4) the social and cultural implications of science. Even a nonvigorous examination of school science programs, he suggested, reveals deficiencies in the coverage of social and cultural implications. O'Hearn recommended that science educators use the idea of "the future" as a strategy. It is "a logical organizing principle with which to develop science literacy. Its adoption as an organizing principle would enhance, not detract, from teaching of science including scientific knowledge, the processes of science, and the nature of science. Clearly, it is closely tied to social imperatives and appears to have great educational potential as a tool for learning. From all of these points of view, it seems not only a rational, but an ideal means to use to increase the literacy in science for future generations, while retaining the strength of the subject content of our present courses" (p. 111). O'Hearn's definition and recommendation make sense, but science educators failed to recognize or to try out either of them. At this point, one might well begin to ask why these, like so many good ideas, remained unimplemented.

In the mid-1970s, Victor Showalter directed a program called *Unified Science Education*. In a program newsletter (1974), Showalter and his colleagues presented a thorough overview of the idea of scientific literacy: "In many ways, scientific literacy represents the goal of a liberal or general education in science. Ideally, each citizen has made and continues to make satisfactory progress toward this goal" (p. 1). The team developing *Unified Science* based their framework on the literature of the previous fifteen years and on discussions with scientists and science educators. They proposed seven "dimensions" of scientific literacy and elaborated on them in some detail. The dimensions included the nature of science, concepts in science, processes of science, values of science, science and society, interest in science, and the skills of science. The authors view these dimensions as a continuum along which individuals progress. If one aspect of one dimension—for example, "tentativeness"—is an aspect of the nature of science, the sophistication and the application of the idea of tentativeness in science increases as a person becomes more scientifically literate. Showalter and his colleagues clarified the rationale for each of the seven dimensions. Under the nature of science, for example, they listed tentative, public, replicable, and so on. This proves a simple, functional place to begin understanding scientific literacy in its various dimensions.

Paul DeHart Hurd (1975) restated his scientific literacy themes in "Science, Technology, and Society: New Goals for Interdisciplinary Science Teaching," an article whose title incorporates the emerging STS theme as well as an interdisciplinary emphasis. Hurd reiterates the theme of teaching science in a personal and social context, but he also emphasizes the place of technology in the science curriculum; values, decision making, and the role of the laboratory in science teaching; and an integrative perspective. Integrative principles, he argues, can serve as conceptual links to the various scientific disciplines: "We have little hope of resolving population, food, health, water, pollution, and many other problems of human concern unless we can relate disciplines and teach them in an integrative mode" (p. 30).

Benjamin Shen (1975) took a social-contextual perspective in discussing practical, civic, and cultural scientific literacy, each of which involves different objectives, audiences, content, format, and delivery. Practical scientific literacy—the most urgent and most neglected—includes "the possession of the kind of scientific and technical knowledge that can be immediately put to use to help solve practical problems. The most basic human needs are health and survival; much of practical science literacy has to do with just those needs" (p. 27).

Clearly, as Shen suggests, because of its global perspective, development must take advantage of mass communications, for example, The Children's Television

Workshop, Dr. Spock's best-selling book on baby care, and satellite communications.

Civic scientific literacy is the cornerstone of informed public policy. Shen refers to "considered common sense opinions": "The aim of civic science literacy is to enable the citizen to become more aware of science and science-related issues so that he and his representatives can bring their common sense to bear upon these issues" (p. 28). He makes two recommendations: first, a greater exposure, in the media and in the schools, to science and its social implications; and second, presenting the science behind science-related public issues in plain language. Once technical jargon is eliminated, the average citizen, using common sense, can make a "considered" decision.

"Cultural science literacy," Shen states, "is motivated by the desire to know something about science as a major human achievement" and cites *Nova* as an example of a television program that develops cultural scientific literacy. He concludes by referring to "ordinary language philosophy"—I would suggest, as an example, Mortimer Adler's work (Adler, 1981, 1984, 1987)—and draws a parallel to science: "What science literacy needs is a planned proliferation of good 'ordinary language science.' Even though this will not contribute to scientific research, the social benefits will be valuable in themselves and well worth the effort" (p. 28).

Benjamin Shen's article provides a unique and valuable perspective in the search for scientific literacy. He describes a social orientation and contextual emphasis for the various topics elaborated by other authors. Shen's view also encompasses a continuum of global, national, and personal dimensions.

In 1976, Milton Pella published "The Place or Function of Science for a Literate Citizenry." In a concerned, even angry, tone, Pella criticized the indiscriminate use of science vocabulary and the proliferation of catch phrases to describe the role, place, and content of science in general education. From his perspective, "a scientifically literate citizenry understands some of the knowledge library of science, knows some of the limitations and potentials of the contents of the library, knows how and when to apply the knowledge theory, knows where the contents of the library come from, and knows some of the regulatory principles involved in knowledge production and use" (p. 99).

Pella recognized the importance of policies and the potential need for national standards: "In the absence of some uniform policy relative to the goals of the country and of the world, it will be extremely difficult to define operationally or to adequately research the topic of 'what is adequate scientific literacy.' Let us recognize that all people cannot understand at the same level. There are some whose knowledge of the natural or man-made universe will be limited to descriptions or classifications of direct or indirect sensory experiences. There are others who will be able to operate with theoretical concepts and laws at the highest quantitative level. There also are countless levels of understanding knowledge in between" (p. 100). In terms of the "library" of scientific knowledge, scientific literacy must accommodate a continuum of understanding.

Table 3.2 compares three views of scientific literacy from the 1970s.

TABLE 3.2 *Some Characteristics of Scientific Literacy: The 1970s*

Michael Agin (1974)	*Victor Showalter (1974)*	*Benjamin Shen (1974)*
1. Science and society	1. Nature of science	1. Practical science literacy
2. Ethics of science	2. Concepts in science	2. Civic science literacy
3. Nature of science	3. Processes of science	3. Cultural science literacy
4. Knowledge of the concepts of science	4. Values of science	
5. Science and technology	5. Science and society	
6. Science and the humanities	6. Interest in science	
	7. Skills associated with science	

The 1980s: A National Goal

During the 1980s, the purpose of science education was clearly linked to the idea of scientific literacy, and the literature on the subject increased substantially. Unfortunately, the term also began to take on a symbolic value distinct from its past conceptual development because individuals used it in a variety of ways.

Compared with its 1971 statement, the NSTA 1982 position statement "Science-Technology-Society: Science Education for the 1980s" suggested an even stronger connection between science and society and reemphasized most of the defining characteristics of scientific literacy from the earlier statement. These two NSTA policies stand in contrast to the NSF programs of the 1960s and 1970s. The interrelationship of science and society, the role of technology, and aspects of science and technology established a perspective on scientific literacy that did not filter down to actual school science programs and practices (Harms and Yager, 1981).[2]

In the early 1980s, reports from organizations outside the science education community brought the issue to national attention. In April 1983, the National Commission on Excellence in Education (NCEE) reported an ominous trend to the American people in *A Nation at Risk: The Imperative for Educational Reform*. The report cited economic- and defense-related risks along with indicators that educators were unable to respond—low student achievement scores, functional illiteracy, and declining test scores. Focusing on science and technology, the report quoted science educators who proclaimed that the nation was "raising a generation of Americans that is scientifically and technologically illiterate" (Paul DeHart Hurd, p. 10) and that there was "a growing chasm between a small scientific and technological elite and citizenry ill-informed, indeed uninformed, on issues with a science component (John Slaughter, p. 10). The NCEE proffered many policy recommendations, including one for science: "The teaching of *science* in high

---

2. Because this section concentrates on a definition and a conceptual framework for scientific literacy, it does not discuss Project Synthesis (Harms and Yager, 1981), although this project and the subsequent report had a significant impact on contemporary science education reform.

school should provide graduates with an introduction to: (a) the concepts, laws, and processes of the physical and biological sciences; (b) the methods of scientific inquiry and reasoning; (c) the applications of scientific knowledge to everyday life; and (d) the social and environmental implications of scientific and technological development. Science courses must be revised and updated for both the college bound and those not intending to go to college" (p. 25).

Recommendations (a) and (b) are similar to the goals that prevailed in many programs developed during the 1960s and 1970s, while recommendations (c) and (d) emphasize general education. In some respects, these goals incorporate ideas expressed in earlier decades, for example, that science and technology education should relate to everyday life and to social and environmental issues. In addition, by recognizing the needs of noncollege bound students, the report reinforced a general education view that science programs should meet the needs of all students. *A Nation at Risk* highlighted scientific literacy as a national priority and helped define its meaning.

Although many reports on education included science and technology, one, *Educating Americans for the 21st Century* issued by the National Science Board (NSB) (1983), addressed them directly and informed a number of policies and programs during the 1980s. Summarizing objectives for science and technology education, the NSB committee stated that "students who have progressed through the nation's school system should be able to use both the knowledge and products of science, mathematics, and technology in their thinking, their lives, and their work. They should be able to make informed choices regarding their own health and lifestyles based on evidence and reasonable personal preferences, after taking into consideration short- and long-term risks and benefits of different decisions. They should also be prepared to make similarly informed choices in the social and political arenas" (p. 45).

According to contemporary articles and reports, the process of rethinking scientific literacy should be grounded in the highest aspirations of our society: citizenship and democratic participation. This broad purpose can be translated into more concrete policies by introducing personal applications such as health and by understanding the relationship of science and technology to social and cultural dimensions. Technology is an essential part of scientific literacy.

A Nation at Risk produced the most immediate and widespread impact, but others also offered important ideas. *High School* by Ernest Boyer (1983) of the Carnegie Foundation for the Advancement of Teaching recommended a two-year science program based in the biological and physical sciences. The study of science should introduce "students to the processes of discovery—what we call the scientific methods" (p. 106). As Boyer elaborates, "These courses should be taught in a way that gives students an understanding of the principles of science that transcend the disciplines. The search for general principles of science can be, if not properly done, a superficial exercise. But, if carefully designed, an interdisciplinary view will give all students—both specialists and non-specialists—a greater understanding of the meaning of science and scientific process" (p. 107). He suggests an integrated curriculum for school science, an approach referred to as a STS theme, and considers the connection to citizenship: "Not all students are budding scientists, but becoming a responsible citizen in the last decade of the twentieth century means that everyone must become scientifically literate. Having a substantial knowledge of scientific facts and processes, and understanding more about the interdependent world in which we live are essential parts of the case of common learning" (p. 107).

In a section entitled "The Impact of Technology," Boyer outlines a plan for a course: "we recommend that all students study technology: the history of man's use of tools, how science and technology have been joined, and the ethical and social issues technology has raised. During this proposed one-semester course, a student might well look at one technological advance—the telephone, the automobile, television, or the microcomputer, for example—trace its development, and examine the positive and negative impact it has had on our lives today" (p. 110).

He further underscores the relationship between technology and society: "The great urgency is not 'computer literacy' but 'technology literacy.' The need for students to see how society is being reshaped by our inventions, just as tools of earlier eras have changed the course of history. The challenge is not learning *how* to use the latest piece of hardware but asking *when* and *why* it should be used" (p. 111).

Ernest Boyer's comments on the contemporary reform of science education adds a significant perspective: science and technology education is for *all* students. Twenty-five years ago, in the era of curriculum reform, science education concentrated on programs for future scientists and engineers. Today we need to direct our attention to all students, especially those for whom twelfth grade will be their last formal experience with science and technology.

In spring 1983, the journal *Daedalus* devoted an issue to the theme of scientific literacy. In his essay, Jon Miller (1983b) considered two different meanings of the term *literate:* to be learned and to be able to read and write. He reviewed relevant empirical studies and described his own empirical research on scientific literacy, which focused on policy. Miller's study considered the public's ability to understand the scientific process, basic science constructs, and science policy issues. According to his findings, only about 10 percent of Americans understood scientific processes, about 50 percent understood basic science constructs, and approximately 41 percent qualified as understanding science policy issues. When Miller devised a single measure incorporating all three dimensions, only 7 percent qualified as scientifically literate.

Miller (1983a) presented a model for science and technology policy formulation in *The American People and Science Policy.* The model uses a pyramid whose base represents what Miller terms the "nonattentative" public; according to his analysis, most of the American public is nonattentive to science policy issues. The next level represents individuals interested in a particular policy area, such as the environment, who are willing to become—and remain—informed in that area. Near the top of the pyramid, and thus representing a very small percentage of people, are policy leaders and decision makers.

In another essay in the *Daedalus* issue, Arnold Arons (1983) described the scientifically literate person and illustrated instructional strategies to help achieve scientific literacy. He suggested concentrating on the "idea first and name afterwards" using a series of general questions, for example, How do we know? Why do we believe? He also recommended covering less material, taking more time, and giving students the opportunity to pursue a small number of major scientific ideas in order to make their knowledge operative rather than declarative: Why do we believe the earth revolves around the sun? Why do we believe that matter is discrete in structure? What is the evidence for the atomic molecular theory? Because his instructional approach incorporates epistemological, philosophical, and historical aspects of science, it assumes a general education and humanistic perspective.

During this same period, however, Morris Shamos began publishing articles in which he argued that scientific literacy for all students was an unrealistic goal (Shamos, 1984, 1988a, 1988b). The articles convey the impression that he was less concerned about scientific literacy than he was critical of science educators and their rhetoric of reform. Shamos (1984) argues for technological literacy on the grounds that technology is an "easier target to hit" because "one does not need to understand the ultimate causes of things to appreciate their end uses" (p. 33). His arguments caused many science educators to reflect on how they used the terms *scientific literacy* and *technologic literacy*.

In the NRC report *Improving Indicators of the Quality of Science and Mathematics Education in Grades K–12,* Richard Murnane and Senta Raizen (1988) developed a statement on scientific literacy. The dimensions of scientific literacy include the nature of the scientific worldview, the nature of the scientific enterprise, scientific habits of mind, and the role of science in human affairs (p. 16). The scientific worldview is based on a set of interconnected ideas about the nature of the natural world. Scientists express these ideas at various levels of complexity and abstraction. There are conceptual schemes, such as evolution; theories and models, such as gravitation; and particular concepts, such as cycles, scale, and energy. The scientific enterprise consists of the ethics and values that define the rules for developing and expressing scientific explanations. Science is theoretical and empirical. The body of knowledge called *science* develops out of a consensus among scientists, it is tentative, and it is represented publicly in presentations and publications.

Scientific habits of mind include the scientific method, the processes of science, and critical thinking. Although no single scientific method exists, scientists consistently use a variety of techniques, methods, and procedures in their work and describe them in their reports. Science in human affairs expresses the general idea that people need to understand something of the connection between science and society. If the scientific world influences society, society also influences science through funding priorities. Cultivating a sense of social and scientific history encourages public appreciation of their connection.

In 1989, the AAAS held a forum on scientific literacy organized by Audrey Champagne and her colleagues, who subsequently edited a small book based on the meeting. In an introductory chapter, Audrey Champagne and Barbara Lovitts (1989b) consider the barriers to consensus. One of the foremost is the confusion of educational purposes, course content, instructional methods, and student outcomes. Three aspects—purposes, means, and outcomes—often intermingle in discussions of scientific literacy.

In preparation for the national forum, the AAAS conducted a survey in which samples of scientists, educators, teachers, and policy analysts were asked about the meaning of scientific literacy. The respondents rated the importance of fifteen attributes (that is, capabilities and attitudes) of scientific literacy for high school graduates. The highest ranked items were

- Read and understand science articles in the newspaper.

- Read and interpret graphs displaying scientific information.

- Engage in a scientifically informed discussion of a contemporary issue.

- Apply scientific information in personal decision making.

- Locate valid scientific information.

Champagne and Lovitts contrasted these with the lowest ranked items, which included defining scientific terms, describing natural phenomena, providing explanations for science concepts, and assessing scientific methodologies. In their discussion of the responses, the authors point out some contradictions in these ratings: how, for example, can an individual read and understand a scientific article in a newspaper if that individual cannot define terms and explain scientific concepts and processes? Respondents probably understood the contradiction, but more important, the analysis suggests the need for a holistic perspective to complement the reductive perspective.

In the AAAS forum volume, Morris Shamos described scientific literacy in the context of science programs for the elementary school. He considers several levels, beginning with E. D. Hirsh's (1987) *Cultural Literacy*. At the next level, Shamos proposes functional scientific literacy, which is characterized by the ability to read, write, and converse coherently about scientific issues. This level goes beyond vocabulary and involves some knowledge. For Shamos, this level displays another interesting characteristic: that is, individuals who are functionally literate in science take a more active position when they discuss science topics and issues. At the third level, true scientific literacy, individuals have a highly sophisticated understanding of the scientific enterprise, including the major conceptual schemes and processes of science. Few people attain this level, and thus, according to Shamos, scientific literacy is not a viable goal for science education. It should be obvious why I do not agree with this conclusion.

In an essay on elementary school science curricula, Angelo Collins (1989) conceives of scientific literacy in terms of the nature of science, specifically the structural (content), procedural (processes), and human components, and identifies a balance among these components that is overlooked in other discussions of scientific literacy (Collins, 1989).

The end of the decade also witnessed the publication of the results of the first phase of *Project 2061, Science for All Americans* (Rutherford and Ahlgren, 1989). This report presented a substantive view of scientific literacy, the knowledge, skills, and attitudes that all students should acquire because of their total school experience:

"*Science for All Americans* is based on the belief that the scientifically literate person is one who is aware that science, mathematics, and technology are interdependent human enterprises with strengths and limitations; understands key concepts and principles of science; is familiar with the natural world and recognizes both its diversity and unity; and uses scientific knowledge and scientific ways of thinking for individual and social purposes" (p. 4).

This report covers a range of topics, many of which are part of most school science programs, but it presents a view of scientific literacy that, while consistent with the discussion in this chapter, contrasts with the traditional view of school science programs. Traditional disciplines, such as physics, chemistry, biology, and geology, for example, are not the organizing basis for scientific concepts and processes. *Science for All Americans* softens these boundaries and emphasizes connections between and among disciplines. In addition, the report emphasizes big ideas rather than discrete vocabulary and considers topics integral to definitions of scientific literacy but not typically introduced in many school science programs— the nature of the scientific enterprise, the history of science and technology, and the major conceptual themes of scientific thinking. The structure of the report forms a continuum for scientific literacy: the set of chapters presents a conceptual framework, the headings within chapters are conceptual categories, the paragraphs are on specific topics, and the vocabulary is contained within each paragraph.

When it was published in 1989, *Science for All Americans* presented one of the most comprehensive and innovative statements of scientific literacy in the history of science education. Although not without some minor weaknesses, such as an overemphasis on knowledge and an underemphasis on inquiry and design abilities, it does represent a major "finding" in the search for scientific literacy. Along with the NRC report (Murnane and Raizen, 1988) and the AAAS report (Champagne and Lovitts, 1989a), it offered a clear statement of the dimensions of scientific literacy and the issues involved in encouraging scientific literacy in American students. I would also note that use of the term *scientific literacy* to refer to the purposes of science education was not confined to the United States. Indeed, leaders in other countries also offered insights and established policies based on this purpose. For example, the Science Council of Canada published *Science Literacy: Towards Balance in Setting Goals for School Science Programs* (Roberts, 1983) and the Israeli Science Teaching Center published *Scientific and Technological Literacy for All: An Israeli Educational Challenge for the Future* (Chen and Novik, 1986). Table 3.3 summarizes some defining characteristics of scientific literacy during the 1980s.

The 1990s: Clarification and Criticism

In the 1990s, scientific literacy has become the single term expressing the purposes of science education. Although some continue to argue about its importance and some have questioned its viability as a goal, others have put standards and benchmarks in place that provide thorough working definitions of scientific literacy. At

TABLE 3.3 *Some Characteristics of Scientific Literacy: The 1980s*

National Science Teachers Association *Science-Technology-Society: Science Education for the 1980s* (National Science Teachers Association, 1982)	National Commission on Excellence in Education *A Nation at Risk* (National Commission on Excellence in Education, 1983)	"Scientific Literacy: A Conceptual and Empirical Review" (Miller, 1983b)	*Improving Indicators of the Quality of Science and Mathematics Education in Grades K–12* (Murnane & Raizen, 1988)	American Association for the Advancement of Science *Science for All Americans* (Rutherford & Ahlgren, 1989)
1. Scientific and technological process and inquiry skills	1. Concepts, laws, and processes of physical and biological sciences	1. Scientific approach	1. The nature of the scientific worldview	1. The nature of science
2. Scientific and technological knowledge	2. Methods of scientific inquiry and reasoning	2. Basic science constructs	2. The nature of the scientific enterprise	2. The nature of mathematics
3. Skills and knowledge of science and technology in personal and social decisions	3. Applications of knowledge to everyday life	3. Science policy issues	3. Scientific habits of mind	3. The nature of technology
4. Attitudes, values, and appreciation of science and technology	4. Social and environmental implications of scientific and technological development		4. Science and human affairs	4. The physical setting
5. Interactions among science-technology-society via context of science-related societal issues				5. The living environment
				6. The human organism
				7. Human society
				8. The designed world
				9. The mathematical world
				10. Historical perspectives
				11. Common themes
				12. Habits of mind

mid-decade, discussion seems to have shifted from purpose and policies to translating those policies into programs and practices. This section does not attempt a complete review of the literature on scientific literacy; rather, it brings together a few statements supporting different perspectives and provides background for Chapter 4.

In January 1990, the Board of Directors of NSTA adopted "Science Teachers Speak Out: The NSTA Lead Paper on Science and Technology Education for the 21st Century." The preamble of this position statement proclaims that major reform must necessarily meet a goal of scientific literacy. Citizens must be prepared to understand and deal rationally with the issues of a scientific and technological world and the opportunities it offers. The affirmation of scientific literacy extended to general discussions of curriculum reform, instructional support, and the preparation of teachers.

In 1991, Robert Hazen and James Trefil (1991a) published *Science Matters: Achieving Scientific Literacy*. They discussed their view of scientific literacy in the popular press and in various professional magazines and presented a rationale for their position: "What non-scientists so need is the background to grasp and deal with matters that involve science and technology. It is this ability to understand science in its day-to-day context that we propose to call scientific literacy. . . . The scientifically literate non-scientists need to understand a little bit of several disciplines to cope with such issues. Scientific literacy thus is a grasp of an eclectic mix of facts, vocabulary, and principles. It is not the specialized knowledge of the experts, nor does it rely on jargon and complex mathematics" (p. 44). Hazen and Trefil present basic scientific facts, vocabulary, and principles in a clear and straightforward manner. Their discussion includes many common and interesting examples. Compared with earlier discussions of scientific literacy, the book offers little about the nature of science, the history of science, the nature of technology, and the role and place of science in society. It outlines scientific principles as follows:

The Most Basic Principle of Science

- The universe is regular, predictable, and quantifiable

The Principles Shared by All Sciences

- Newton's laws governing force and motion

- The laws of thermodynamics governing energy and entropy

- The equivalence of electricity and magnetism

- The atomic structure of matter

Principles for Specific Natural Systems Combined with the Basic Principles

- Galaxies

- Stars

- Earth
- Living things

Basic Principles for the Earth Sciences

- Plate tectonics
- Earth cycles

Basic Principles for the Life Sciences

- All living systems are based on chemistry
- All living systems are made of cells
- All living systems use the same genetic code
- All living systems evolve by natural selection
- All living systems are interconnected

During the 1990s articles and books critical of the goal of scientific literacy appeared. Edgar Jenkins (1990a, 1992) concluded that adopting the goal of scientific literacy would most likely burden school systems beyond their capacity to adapt. His analysis on the multidimensional nature of scientific literacy is one of the more insightful, but when he extends it to discuss its implications for curriculum, instruction, and assessment, he demonstrates the burden of implementation.

Other articles also critiqued the relationship of school science to practical action (Layton, 1991); the priority of different goals, such as practical reasoning, habits of mind, and science as a product of human thought and action (Atkin and Helms, 1993); the narrow confines of the idea (Eisenhart, Finkel, and Marion, 1996); and the historical context and contemporary conflict between science as an elitist enterprise and education as a populist enterprise (Raizen, 1991). In the most popular critique, *The Myth of Scientific Literacy,* Morris Shamos (1995) argued that scientific literacy is an unachievable goal and that pursuing it misuses talent and resources.

The major statements on scientific literacy in mid-decade were *Benchmarks for Science Literacy,* developed by the AAAS's *Project 2061* and published in 1993, and the *National Science Education Standards,* developed by the NRC and published in 1996, both of which are discussed in greater detail in later chapters. Trefil and Hazen (1991b) summarize the characteristics of scientific literacy in the 1990s:

1. knowing
2. energy
3. electricity and magnetism
4. the atom
5. the world of the quantum

10. astronomy
11. the cosmos
12. relativity
13. the restless earth
14. earth cycles

6. chemical bombing

7. atomic architecture

8. nuclear physics

9. particle physics

15. the ladder of life

16. the code of life

17. evolution

18. ecosystems

Conclusion

The search for scientific literacy has led us in a variety of directions over a relatively short period. Although I cannot claim that science educators have developed a consensus on scientific literacy, neither can I say, as many do, that no one stops to define the terms. It is also the case that the term has taken on the characteristic of a slogan, since many use it without bothering to define what they mean by it or to reveal whose definition they are using. Still, it would seem that there is more consensus than science educators acknowledge. Could it be that few have stopped to find out what others have said? I think this is so. This review incorporates a number of individual statements and reports, but it is hardly exhaustive. For that I would refer readers to *Scientific and Technological Literacy, Meanings and Rationales: An Annotated Bibliography* (Layton, Jenkins, and Donnelly, 1993).

This search for scientific literacy does, however, yield several conclusions:

- Scientific literacy is a metaphor referring to the purpose of science education.

- Scientific literacy emphasizes a general education orientation.

- Scientific literacy expresses norms or standards for science education programs, methods, and assessments.

- Scientific literacy illustrates different perspectives in science education.

- Scientific literacy represents a continuum of understandings.

- Scientific literacy incorporates multiple dimensions.

- Scientific literacy includes both science and technology.

4

Defining Scientific Literacy

The phrase "scientific literacy for all learners" expresses the major goal of science education—to attain society's aspirations and advance individual development within the context of science and technology. Certainly, most science educators rally around this statement because it embodies the highest and most admirable goals of science teaching. But what does it actually mean?

Educational Definitions

In *The Language of Education,* Israel Scheffler (1960) points out that a general definition may be *stipulative* or *inventive.* An individual may stipulate that a certain term, such as *scientific literacy,* is equivalent to some other term or description. I have stipulated, for example, that scientific literacy encompasses the major purposes of science education—scientific knowledge, methods, and careers, as well as personal development and social aspirations (Bybee and DeBoer, 1993; DeBoer and Bybee, 1995). In contrast, an individual who stipulates a definition without setting an equivalent invents a unique definition. Benjamin Shen's (1975) definition of scientific literacy as practical, civic, and cultural is an inventive definition. Other examples include *Science for All Americans* (Rutherford and Ahlgren, 1989), the book is based on the work of *Project 2061,* which defines the essential understandings of a scientifically literate adult, and the *National Science Education Standards* (National Research Council, 1996), the document developed by the NRC, which describes what all students should know and be able to do after thirteen years of science education. Scheffler describes yet another type of educational definition, what he calls a *descriptive* definition, which serves some of the same functions as stipulative definitions and explains the defined term by giving an account of its previous usage. The original work of Milton Pella and his colleagues (Pella, O'Hearn, and Gale, 1966) developed a descriptive definition of scientific literacy based on hundreds of references to the term.

Finally, Scheffler describes *programmatic* definitions, which imply an active moral dimension. A stipulative definition of scientific literacy which then asserts that science teachers ought to design curricula and use certain strategies to develop scientific literacy goes beyond mere stipulation: it implies a dimension of choice, decision making, and action. Some of Pella's and O'Hearn's (1976) later arguments about recommendations for using alternative futures as the context for scientific literacy fall into this category. When individuals or committees define scientific literacy as the experience students should have (National Science Teachers Association, 1971), they present a programmatic definition. They imply that actions ought to be taken based on the definition.

Scheffler summarizes the three types of definitions: "The interest of stipulative definitions is communicatory, that is to say, they are offered in the hope of facilitating discourse; the interest of descriptive definitions is explanatory, that is, they purport to clarify the normal application of terms; the interest of programmatic definitions is moral, that is, they are intended to embody programs of action" (p. 22).

Claiming that scientific literacy has not been defined makes for engaging rhetoric, but the assertion is wrong (see Chapter 3). It betrays a glaring lack of historical awareness. In many cases, the issue of accepting and using the term *scientific literacy* boils down to personal attitudes—what individuals like or don't like—and that is what ultimately determines its acceptance or rejection. It is much easier to use the term as a slogan and claim that it is seldom defined.

Defining Literacy

While discussing the larger issue of definitions, a colleague suggested that the explanation of scientific literacy should begin with the term *literacy*—what it means and how it is evaluated. This recommendation is based on the commonly held assumption that literacy means the ability to read and write. But the measure of literacy has varied over time, from being able to write one's name to being highly educated (Graff, 1986). Literacy has also been defined by the ability to read scripture or newspapers. The 1940 U.S. Census defined it as having completed the fourth grade (deCastell, Luke, and MacLennar, 1986). Although the definitions have varied, being literate has consistently referred to mastering the processes needed to interpret culturally significant information (deCastell and Luke, 1986). Several other observations about literacy also apply to this discussion. First, the acceptable levels of literacy have increased over the years. Second, the number of those within the population who are expected to be literate has expanded to include virtually everyone except the severely or developmentally disabled. And third, schooling is not synonymous with acquiring literacy.

Lawrence Cremin's (1988) discussion of *inert literacy* as opposed to *liberating literacy* provides a useful insight. *Inert literacy* refers to the ability to sign a document or read a passage. In contrast, according to Cremin, "Liberating literacy in metropolitan America meant that people could read freely and widely in search of whatever information and knowledge they chose. It also meant that literacy could serve a multitude of purposes in a multitude of ways" (p. 658). Liberating literacy provides access to materials that open people's minds and introduce them to new ideas and aspirations. Isn't this what we wish for all students enrolled in science classes?

Scientific Literacy as a Slogan

"Scientific literacy" certainly qualifies as an educational slogan. As we have seen, science educators use the phrase repeatedly and with assurance, and see little need to clarify their meaning. According to Scheffler (1960), "Slogans in education provide rallying symbols of the key ideas and attitudes of educational movements. They both express and foster community spirit, attracting new adherents and providing reassurance and strength to veterans" (p. 36). A slogan has a positive value among educators. Scientific literacy becomes a rallying cry for contemporary reform; it serves to unite science educators behind a single statement representing the purposes of science education. In time, individuals begin to interpret slogans in a variety of ways. So, for example, scientific literacy means whatever the speaker intends and the listener hears, whether the speaker clarifies the intended meaning or not. More often than not, analysis reveals differences between intention and interpretation. Purpose statements, discussed in earlier chapters, are able to unify individuals in a cause; in order to do so, however, they have to be abstract and not clearly defined.

Scientific Literacy as a Metaphor

In his discussion of educational metaphors, Scheffler comments, "Metaphorical statements often express significant and surprising truths, unlike stipulations which express no truths at all, and unlike descriptive definitions, which normally fail to surprise. Though frequently like programmatic definitions conveying programs, metaphors do so always by suggesting some objective analogy, purporting to state truths discovered in the phenomena before us" (p. 47). As a figure of speech, scientific literacy serves as a metaphor for the goals and purposes of science education. The traditional sense of literacy, being able to read and write, assumes another meaning. Scientific literacy refers to being well-educated and well-informed in science. It does not mean being able to read and write about science or being able to understand science vocabulary. Rather, it echoes Edward Hirsh's (1987) approach in *Cultural Literacy,* which he defines as possessing the basic information needed to thrive in the modern world: "To be truly literate, citizens must be able to grasp the meaning of any piece of writing addressed to the general reader" (p. 12). He means reading newspapers of substance and, I would infer, magazines of substance; but here the idea of literacy broadens in response to the question, What about watching television programs of substance?

What does Hirsh have to say about scientific literacy? As it turns out, James Trefil, a professor of physics, developed the list of science terms in *Cultural Literacy.* In a note at the end of the book, Trefil describes his rationale: the term must be "truly essential to a broad grasp of a major science. This criterion may not assure true scientific literacy, but it should at least help overcome serious illiteracy" (p. 148). He implies the existence of a continuum with *serious illiteracy* at one end and *true scientific literacy* at the other. "The goals of scientific literacy," he says, "are no different from those of other kinds of cultural literacy: to provide each citizen with the basic framework of knowledge into which public debate can be placed" (p. 149).

In the cultural context, Hirsh and Trefil equate scientific literacy with the ability to read and use common scientific terms, such as *acid, amoeba, atom, buffer, cyclotron, diffusion, electron, fauna, galaxy, homeostasis, igneous rock, jetstream, kinetic energy, laser, magnetism, noble gas, ohms, particle accelerator, quark, recessive trait, shale, technology, ultraviolet, vaccine, water table, X-ray, Y chromosome,* and *zodiac,* all of which appear in the *Cultural Literacy* list. The term *literacy* and its opposite, *illiteracy,* are commonly used in the context of a minimal or functional level of competence. Scientific illiteracy could thus be defined as being unable to read or use any of the terms listed, and functional scientific literacy as the ability to read and use these terms. But scientific literacy should not be defined in this narrow sense, as the lowest level of functional literacy. Most would agree that we want—and expect—more than functional literacy from thirteen years of formal science education.

The literacy metaphor also provides a way of thinking about effective science teaching. Most educators agree that the best way to learn and become truly literate in a foreign language is by immersing oneself in the culture of that language: to learn French, live in France for a year. In every culture, language acquires dialects, colloquialisms, and idiomatic subtleties not conveyed in language classes. This is also true of science and technology. Physics and biology, for example, differ conceptually but they also embrace different worldviews and different methodologies, as do the cultures of science and engineering. Viewed metaphorically, scientific literacy could imply immersing students in the authentic questions and processes of science, and in the real problems and design strategies of technology as opposed to the meaningless memorization vocabulary. Immersion of this kind would be complemented with an introduction to the history and nature of science and technology (Bybee et al., 1992a; Jenkins, 1990b). Certainly, scientific literacy includes understanding the meaning of scientific terminology, but it is more than mere vocabulary. It extends to concepts (Hazen and Trefil, 1991). Scientific literacy suggests a goal for *all* students, immersion in a culture, standards of achievement, and learning more than scientific and technological vocabulary.

Toward a Definition of Scientific Literacy

The "dimensions" of scientific literacy (see Chapter 3) refer to aspects of a programmatic definition. School science programs and classroom practices should reflect these dimensions.

Scientific Literacy Clarifies the General Education Purposes of Science Education

The tension between scientists and educators over the appropriate emphasis and content of science education is ongoing. Scientists, whose views have been shaped by their professional education and orientation, emphasize science content, often disregarding the educator's perspective. Conversely, educators, also influenced by their professional experiences, tend to overemphasize educational issues at the expense of science content. Senta Raizen (1991) has characterized these positions as elitist (scientists) versus populist (educators). Philip Johnson (1962) also described this tension, so it is not unique to contemporary reform. Both miss the point of general education, the position that is implied by the term *scientific literacy*.

General Education in a Free Society (Harvard Committee, 1945) defined both general and special education: General education "is used to indicate that part of a student's whole education which looks first of all to his life as a responsible human being and citizen; while the term, special education, indicates that part which looks to the student's competence in some occupation" (p. 51). General education represents an orientation to curriculum that values the experience and learning common to all students. Its intent is to introduce students to those experiences, relationships, and ethical matters that are common to all members of a society at a particular historical moment. As the passage from *General Education in a Free Society* indicates, citizenship is a controlling factor. Designing science curricula should begin not by outlining the content associated with the structure of the discipline but by asking what it is a student ought to know, value, and do as a citizen.

When are we, in fact, engaged in general, as opposed to special, education in science? This question often relates to students interested in pursuing science as a special career, or to recruiting students into the professional science pipeline, which would argue for a variety of precollege activities, classes, and programs. For clarification, we can again refer to *General Education in a Free Society,* for whom the answer was clear and unequivocal: "Below the college level, virtually all science teaching should be devoted to general education" (p. 156).

The contemporary idea of teaching science from a general education perspective is not new. Morris Shamos (1963), Milton Pella (1967), Victor Showalter (1974), Ernest Boyer (1983), and *Science for All Americans* (Rutherford and Ahlgren, 1989) echo the statements of the Harvard Committee (1945). In 1972, F. James Rutherford published "A Humanistic Approach to Science Teaching" in which he argued for a general education approach. His design for a humanistically oriented science course would connect the sciences with the content and values of the field of history, philosophy, literature, and the fine arts. A humanistic approach to science teaching makes sense for several reasons. First, science shares many of the intellectual, conceptual, imaginative, and aesthetic characteristics attributed to the humanities. Second, scientists influence and are influenced by the history, art, philosophy, and literature of their period. And third, each of the sciences and humanities has its own value and integrity, and all are necessary to society.

In 1988, John A. Moore published "Teaching the Sciences as Liberal Arts—Which, of Course, They Are." After reviewing current programs (K–16), Moore concludes that "the basic responsibility of science teachers in all grade levels is to ensure that as many students as possible understand science and technology to a degree that will make them feel at home in the modern world and enable them to make informed decisions about important questions that deal with scientific matters" (p. 445). This simple, yet comprehensive, statement can serve as a basis for designing science education programs. Moore recommends integrating subject matter into themes, such as the natural world, health, technology, environment, the nature of science, and ethics, and he reiterates many of these same themes in "Cultural and Scientific Literacy" (1995).

In *A Quest for Common Learning,* Ernest Boyer and Arthur Levine (1981) point out that there have been several revivals of general education since the turn of the century, all expressing a recurrent theme: shared values, shared responsibilities, shared governance, a shared heritage, and a shared global vision. Those who have argued for general education have been concerned that society is losing cohesion and splintering into individual factions, each faction pursuing a single interest. This same historical argument has been offered for science education as support for the STS theme and for the 1989 AAAS report *Science for All Americans* (Hlebowith and Hudson, 1991).

The search for scientific literacy is best served by a general education perspective, as a number of science educators would agree (Hurd, 1970; Layton, 1973; Showalter, 1974; Hlebowith and Hudson, 1991; Fensham, 1986/87). Science education programs should be based in a general education approach. This recommendation is supported by the literature on scientific literacy, the current emphasis on providing science education that is appropriate for all students, and emerging global interdependence. But there is an even greater urgency. In *The Disuniting of America,* where Arthur Schlesinger (1992) argues that, at a time when this country needs greater unity and cohesion we are becoming fragmented and divided. To be sure, social problems are multidimensional and will not be solved by education alone, but uniting under the banner of general education, in particular in the sciences, could make a positive contribution.

Scientific Literacy Implies the Same Standards for All Students

As soon as educators had proclaimed that science was for *all* Americans and identified scientific literacy with general education, they faced a new task: clarifying what *science for all* means for curriculum programs and teaching practices. Peter Fensham, an Australian science educator, has written several thoughtful essays on the topic (1985, 1986/87, 1988). He clearly differentiates between teaching science to produce future scientists and engineers and teaching science to produce a scientifically literate citizenry. In many respects, these goals are contradictory, not complementary. Peter Fensham (1985) provided a useful metaphor to help conceptualize the difference between a science program for future scientists and one for all students. The former views science from the outside, from the perspective of someone in society.

In examining the proposition *science for all,* we can begin with the assumption that *all* means what the common definition would suggest, namely, the same basic curriculum and learning outcomes for every student. It does not mean different objectives and different programs for particular individuals or groups of students. It does not mean a set of higher and lower, or harder and easier, standards and experiences. Science for all implies the same standards for all, without exception.

Previous *standards* for school science did not achieve the goals that schools in a democratic society should seek—an education of a quality that allows every student to continue learning as an adult, either formally in colleges and universities or informally. For lower-track students (often underrepresented in science and technology and underprivileged in society), the standards are low, the science education inadequate, and the results discriminatory. The other, higher track, also pursued by a minority of students, is higher only because it is more difficult and advanced, not because its objectives form a core education for all.

Achieving *science for all* involves three goals: the first is related to personal growth and continued development, the second to an individual's role as a citizen in a democratic society and preparation for these duties, and the third to the need to earn a living in one (and probably more) occupation(s) during a lifetime. Thirteen years of science education should develop *basic* skills common to all occupations in our society.

The kind of science education implied by *science for all* is general and liberal rather than specific and vocational. In view of the latter and the current interest in school-to-work programs, science education standards should require that all students be introduced to science-related vocations and prepared for continued study for future careers. Individuals share significant relationships, in this case scientific and technological, with others in the community. General education enhances these connections, encourages cooperation, and affirms the qualities that unite us. Science and technology are part of our shared cultural heritage, a shared set of problems and possibilities relating to science and technology, and a shared future in which science and technology will have a significant presence.

Chapter 3 showed that nearly every perspective on scientific literacy implies a set of normative standards, the idea that scientific literacy could, or ought to be, used as a measure of accomplishment, as a criterion of achievement, and as a measure of program adequacy. The idea of scientific literacy could also be used to structure science programs and in some cases as specifications for various components of those programs.

What are the implications of a *science for all* perspective in science programs? Using Peter Fensham (1985) as a guide, I can summarize by saying that the curriculum design should develop scientific knowledge, skills, and values in a context that affirms a social, rather than an exclusively scientific perspective. The following list presents some of these design criteria:

- Content has immediate and obvious personal and social meaning to the student; it extends and elaborates the student's previous experience, understanding, and skill.

- Content may be challenging, but it must also be achievable at some level by all students.

- Content should be presented in a context that enhances personal and social decision making.

- Content should be taught in a variety of ways that allow the student to demonstrate understanding, skills, and values.

- Content should be assessed in terms of the student's previous knowledge, current achievement, and future potential.

- Content should be assessed in the context of personal and social decision making.

Scientific Literacy Illustrates Different Emphases in Science Education

Most authors and committees seem to have their own distinct view of the educational experiences that constitute scientific literacy and of how a scientifically literate person thinks and acts (Champagne and Lovitts, 1989b; Fourez, 1989). Table 4.1 displays examples of four perspectives. The categories of content, context, experiences, and outcomes summarize varying emphases. I have deliberately drawn my examples from the 1970s since they have merit but are not usually cited in the contemporary literature. There does not seem to be as much disagreement on the general idea of scientific literacy as there is a lack of agreement on terms and emphasis. These four perspectives are not arguing against each other as much as they are ignoring and talking past each other.

In 1982, Douglas Roberts introduced science educators to the idea of "curriculum emphasis," the orientation, or flavor, of a curriculum. The purpose behind science education—the point of learning the material—is conveyed in the structure, organization, and emphasis of policy statements, curriculum materials, and teaching strategies. Roberts (1982, 1983, 1988) provides examples: everyday coping, structures of science, science-technology decisions, scientific skill development, correct explanations, self as explainer, and solid foundation (see Table 4.2).

I have found Roberts's discussion to be extremely useful in designing curriculum, and in analyzing policies and programs. Yet those who are involved in design and development are often unaware of the particular curriculum emphasis they are advocating. Many, for example, focus on the structures of science disciplines or correct explanations without regard for any other emphasis, even passing judgment that any other emphasis is wrong. Subsequent policy statements and curriculum materials then tend to reflect only one orientation. But different curriculum emphases offer wider horizons for scientific literacy. For the most part, science educators have not fully realized the potential of Roberts's powerful idea.

TABLE 4.1 *Perspectives of Scientific Literacy*

Content (Agin, 1974)	*Context (Shen, 1975)*	*Experiences (National Science Teachers Association, 1971)*	*Outcomes (Showalter, 1974)*
1. Science and society	1. Practical scientific literacy	1. Every individual has an opportunity to encounter many science experiences every year	1. The scientifically literate person understands the nature of scientific knowledge
2. Ethics of science	2. Civic scientific literacy	2. Every science teacher should have adequate science facilities, equipment, and supplies and time to use them	2. The scientifically literate person accurately applies appropriate science concepts, principles, laws, and theories in interacting with his or her universe
3. Nature of science	3. Cultural scientific literacy	3. Science should be taught as a unifed discipline, integrated and/or coordinated with other disciplines, such as mathematics, social science, economics, and political science	3. The scientifically literate person uses processes of science in solving problems, making decisions, and furthering his understanding of the universe
4. Concepts of science		4. Increasing emphasis should be placed on science processes, conceptual schemes, and values and less emphasis on factual information	4. The scientifically literate person interacts with the various aspects of his or her universe in a way that is consistent with the values that underlie science
5. Science and technology		5. Direct experiences with the natural world or in the laboratory should constitute the major portion of the science program	5. The scientifically literate person understands and appreciates the joint enterprise of science and technology and their interrelationships with each other and with other aspects of society

TABLE 4.1 *Perspectives of Scientific Literacy (continued)*

Content (Agin, 1974)	*Context (Shen, 1975)*	*Experiences (National Science Teachers Association, 1971)*	*Outcomes (Showalter, 1974)*
6. Science and the humanities		6. Textbooks should facilitate inquiry, not replace laboratory experiences. The use of recorded material (other media as well as printed material) should be an integral part of and depend on laboratory experiences 7. Science education programs should include environmental education that integrates natural phenomena, environmental influences, science, technology, the social implications of science and technology, and economic considerations 8. Opportunities for professional growth should be considered an integral part of science education programs, so that teachers can bring their own deeper insights to bear on the science programs designed for scientific literacy	6. The scientifically literate person has developed a richer, more satisfying and more exciting view of the universe as a result of his science education and continues to extend this education throughout his life 7. The scientifically literate person has developed numerous manipulative skills associated with science and technology

TABLE 4.2 *Curriculum Emphasis in Science Education*

Curriculum Emphasis	View of Science	View of Student	View of Teacher	View of Society
Everyday coping	A knowledge system necessary for understanding and controlling everyday objects, events, and organisms	Must master knowledge to control and to explain natural world	Can explain natural and designed world in terms of scientific and technologic concepts	Individuals must be autonomous and manage personal and social affairs
Structures of science	A system of conceptual schemes that explains objects, organisms, and events and is cumulative and self-correcting	Must understand the conceptual and procedural aspects of science	Must understand the conceptual schemes	Society benefits from scientists who understand the conceptual and procedural system of science and technology
Science–technology decisions	An extension of the human desire to control the environment through technology and directly related to societal issues	Must become an intelligent citizen who makes decisions about science and technology and societal issues	Must present the complex relationships among science, technology, and society	Society needs citizens who will improve it through informed decisions about science-related issues

TABLE 4.2 *Curriculum Emphasis in Science Education (continued)*

Curriculum Emphasis	View of Science	View of Student	View of Teacher	View of Society
Scientific skill development	Scientific explanations result from the correct use of inquiry processes	Must experience and become competent at scientific inquiry	Must provide students with experiences with scientific inquiry	Society benefits from individuals who think and reason like scientists
Correct explanations	Best meaning system for truth of natural objects and events	Must change current conceptions to scientific concepts	Responsible for facilitating conceptual change in students	Society is best served by those who share correct scientific concepts
Self as explainer	Conceptual system influenced by historical moment, concepts, and personal intention	Must have intellectual freedom gained by understanding the influences on scientific thought	Committed to liberal education as basis for exposing what is known	Society needs liberally educated individuals
Solid foundation	A vast and complex system that takes many years to master	Must have a good beginning for a life of scientific study	Responsible for developing the most talented students into future scientists	Society benefits from scientists

Scientific Literacy Represents a Continuum of Understandings and Abilities

Scientific literacy is a continuous process. Referring to the seven outcomes he described, Victor Showalter (1974) noted that "the dimensions and associated factors should be viewed as a specific continuum along which an individual can make progress. This is in contrast to the view that educational objectives are either achieved or not" (p. 1). Milton Pella (1976), in a strong statement on the place and function of science in society, reminded educators that not all individuals can understand science and technology at the same level. Some will understand only concrete experiences, while others will be able to rely on theoretical and quantitative positions in explaining natural phenomena. Many discussions of scientific literacy tacitly suggest an ideological perspective. The unexpressed position is that scientific literacy consists of a set of characteristics—a type of knowledge, understandings, values, abilities, and sensibilities—that can be described, assessed, and thus achieved in a certain space and over a certain time. This position is understandable if one defines scientific literacy as, for example, vocabulary, which allows direct teaching and easy assessment, but most agree that scientific literacy involves much more than vocabulary.

One dimension of the continuum is breadth, which ranges from vocabulary to major conceptual ideas to a contextual understanding of science and technology. Another is depth, an increasingly subtle and sophisticated understanding of scientific concepts such as the atom, evolution, or conservation; scientific inquiry such as questioning, hypothesizing, and explaining; and scientific values such as verification, logic, and tentativeness. Individuals learn over a lifetime and at any one time may be at different points on the continuum in respect to particular topics. Hazen and Trefil (1991a) cite examples of geophysicists who do not know about RNA and DNA and of a Nobel Prize-winning chemist who had never heard of plate tectonics. But no one would doubt the ability of these individuals to learn about RNA, DNA, or plate tectonics if they needed or wanted to do so. There are many examples of individuals who have learned a great deal about a specific topic in a short time (Layton, 1991). For instance, many individuals who have medical problems develop an understanding of their ailment, often at a level comparable to health care professionals. But their understanding is limited to that specific disease, and does not encompass a broad understanding of health, medicine, science, and technology.

Scientific Literacy Incorporates Multiple Dimensions

In his discussion of the difference between memorizing definitions and communicating scientific ideas to another person, Morris Shamos (1989) pointed out the passive-to-active dimension of science literacy. His use of passive-to-active is similar to Cremin's (1988) distinction between "inert" and "liberating" literacy. Benjamin Shen (1975) described personal, civic, and cultural dimensions, and I have described (Bybee, 1993) local-to-global dimensions. David Layton (1973) has discussed the internal view of science, for example, focusing on scientific terms, and the external view of science, for example, discussing science from a social perspective.

Scientific Literacy Includes Both Science and Technology

Many researchers (Hurd, 1975; Pella, 1967; Shamos, 1963; Fleming, 1989; Boyer, 1983; Showalter, 1974) and significant reports, such as *Science for All Americans* (Rutherford and Ahlgren, 1989) and the *National Science Education Standards* (National Research Council, 1996), have included science and technology in their vision of science education. Hurd (1975) states, "Technological literacy ranks with science literacy as a major goal of science teaching" (p. 28). In *Science for All Americans* entire chapters focus on the nature of technology and the designed world, and national standards have been established for science and technology, yet, most science programs give little, and often incorrect, recognition of technology.

Toward a Framework for Scientific Literacy

The following discussion presents a view of scientific literacy that incorporates, but differs from, many of the perspectives described in Chapter 3. The perspective responds to the perceived inadequacy of other perspectives, the apparent discrepancies of other definitions, and the general need to have a comprehensive, yet developmentally appropriate view of scientific literacy especially for science teachers K–12.

Since first reading about the scientists in one discipline who did not know of a major theory in another discipline (Hazen and Trefil, 1991a) and about the Harvard graduates who could not explain why it was hotter in the summer than in the winter, I have been puzzled. Certainly, scientists and Harvard graduates understand more about science than many other members of society, and it is clear that no one possesses total scientific knowledge. At the same time, there are students who, because of a personal interest or a hobby, know a great deal about one scientific subject, and adults who, because of personal circumstances such as a disabled child or an illness (Layton, 1988), become experts in a specific area. Indeed, they can be judged as scientifically literate in that one area, if not in general.

The framework I am proposing is a threshold model, which assumes that scientific and technological literacy are continuously distributed within the population. At one extreme are a small number of scientifically and technologically illiterate individuals, and across the population, a distribution of individuals who demonstrate increasingly greater degrees of scientific literacy. At the other are a small number of individuals whose level of scientific literacy is extremely high. The degree of scientific and technological literacy demonstrated by any individual at any one time is a function of a range of factors—age, developmental stage, life experiences, and quality of science education, which includes an individual's formal, informal, and incidental learning experiences. The model describes certain thresholds that identify degrees of scientific literacy. The framework likewise provides a larger model that is useful to those constructing school science programs or teaching science.

The framework model also favors inclusion rather than exclusion. Some attempts to define scientific literacy assume an either/or perspective: one has it or one does not. A more productive definition recognizes that scientific and technological literacy develop over a lifetime.

The framework I propose also accommodates the fact that a person may, at any time, be compared to the population as a whole and may demonstrate several levels of literacy at once depending on the context, the issue, and the topic. Likewise, subgroups of similar individuals, whether scientists or middle school students, may be located at different points on the scientific literacy continuum. I would note that standards define content and thresholds for scientific literacy.

Illiteracy

Some individuals may be defined as scientifically and technologically illiterate, because of age, stage of development, or developmental disabilities; the percentage of the total population is, however, quite small. If they are asked a question relating to science or technology, for example, they do not have the cognitive capacity to understand the question itself or to locate it within the domain of science or technology. Those responsible for science education usually understand that some people function at this level.

Nominal Literacy

The term *nominal* means existing in name only, so that someone who is "nominally literate" in science may understand that a term, question, or topic is scientific but know little else about it. At this level individuals demonstrate a token understanding of phenomena. Cognitive psychologists would call this *naive theory* and *misconception* (McGilly, 1994; Driver et al., 1994; Bruer, 1993).

Functional Scientific and Technological Literacy

Functionally literate individuals can use scientific and technological vocabulary, but only within a specific context, such as defining a term on a test, reading a newspaper, or listening to a television program. Their knowledge generally lacks the conceptual embellishment of the disciplines and consists of memorized lists of terminology.

Conceptual and Procedural Literacy

Conceptual and procedural literacy means understanding how the concepts of a discipline relate to the discipline as a whole and to the methods and processes of inquiry. In biology, for example, evolution is the unifying conceptual scheme that brings together energetics, genetic continuity, and structure and function. In technology, it is design. General principles such as the laws of thermodynamics and

Newton's laws relate to all scientific disciplines and technologic endeavors. Procedural knowledge and skills, such as the processes of scientific inquiry and technological problem solving, are also relevant. Here individuals actually understand and can use ideas such as *observation* and *hypothesis,* or *optimization* and *constraints* in laboratory investigations or discussions of scientific experiments and engineering developments.

Multidimensional Literacy

Multidimensional literacy recognizes the importance of integral, contextual perspectives. *Integral* means essential for completion, necessary to the whole, and joining with something else. Where the emphasis at the conceptual and procedural level was on scientific disciplines and technological domains, this level of literacy includes the philosophical, historical, and social dimensions of the disciplines: the individual develops an understanding of and appreciation for science and technology as cultural enterprises, making connections within the science disciplines, between science and technology, and between science and technology and larger social problems and aspirations.

A number of researchers have presented frameworks for scientific literacy that recognize what I am calling multidimensional scientific literacy (Pella, O'Hearn, and Gale, 1966; Agin and Pella, 1972; Agin, 1974; Showalter, 1974; Murnane and Raizen, 1988; American Association for the Advancement of Science, 1993; National Research Council, 1996). The dominant example in contemporary reform is *National Science Education Standards.* In this work, the NRC's working groups of scientists, science teachers, and science educators outlined what all Americans should know by age eighteen, but what is more important to this discussion, they also presented an integral and contextual approach to scientific literacy.

The framework I am proposing presents scientific and technological literacy as a continuum in which an individual develops greater and more sophisticated understanding of science and technology. This framework functions as a taxonomy for extant programs and practices and as a guide for curriculum and instruction.

Nominal Scientific and Technological Literacy

In *nominal* literacy, the individual associates names with a general area of science and technology. However, the association may represent a misconception, naive theory, or inaccurate concept. Using the basic definition of nominal, the relationship between science and technology terms and acceptable definitions is small and insignificant. At best, students demonstrate only a token understanding of science concepts, one that bears little or no relationship to real understanding.

Functional Scientific and Technological Literacy

Individuals demonstrating *functional* level of literacy respond adequately and appropriately to vocabulary associated with science and technology. They meet minimum standards of literacy as it is usually understood; that is, they can read and write passages with simple scientific and technological vocabulary. Individuals may also associate vocabulary with larger conceptual schemes—for example, that genetics is associated with variation within a species and variation is associated with evolution—but have a token understanding of these associations.

Conceptual and Procedural Scientific and Technological Literacy

Conceptual and procedural literacy occurs when individuals demonstrate an understanding of both the parts and the whole of science and technology as disciplines. The individual can identify the way the parts form a whole vis à vis major conceptual schemes, and the way new explanations and inventions develop vis à vis the processes of science and technology. At this level, individuals understand the structure of disciplines and the procedures for developing new knowledge and techniques.

Multidimensional Scientific and Technological Literacy

Multidimensional literacy consists of understanding the essential conceptual structures of science and technology as well as the features that make that understanding more complete, for example, the history and nature of science. In addition, individuals at this level understand the relationship of disciplines to the whole of science and technology and to society.

A Discussion of the Framework

This framework is, I would admit, complex and comprehensive. No one could possibly achieve full scientific and technological literacy. But I would not conclude that the framework is useless or that scientific literacy does not comprise the purpose of science education. Developing this kind of literacy is a lifetime task. Some will develop further than others at all levels or within one, depending on their motivation, interests, and experiences: bird watchers, for example, may be more scientifically literate than physicists in classification and habitats of birds; science teachers may enjoy a broader scientific literacy than some scientists and engineers.

Science educators will see the horizontal as well as the vertical development the framework represents. Horizontal development, for example, might entail learning more vocabulary and developing a higher level of functional literacy, but it would not necessarily include developing a greater understanding of the conceptual and procedural nature of scientific disciplines and technological developments.

Although there seems to be a logical and natural progression within the framework, individuals do not necessarily progress through the domains I have described. Cognitive development varies, and an individual may be at different places within the framework at any time and for any given topic. Reason, developmental psychology, and research in the cognitive sciences, suggest, however, that younger, less experienced individuals will tend to be at the nominal and functional levels, while older, more educated scientists, engineers, science educators, and science teachers will be at the conceptual, procedural, and multidimensional levels.

Using this framework for curriculum and instruction does not mandate the development of materials for the functional level of literacy. Programs might incorporate a variety of curriculum (Roberts, 1983), all of which can be based on this framework. The primary challenge is to develop a program that recognizes and enhances learning at all levels of literacy while acknowledging the constraints of students' personal development and interests.

Conclusion

Many in the science education community have used the term scientific literacy, and science educators have argued for programmatic definitions. Yet a clear definition of scientific literacy has not been generally *accepted*. Even *Science for All Americans* and the *National Science Education Standards,* reports that comprise the most complete statements of scientific literacy, are used selectively. Educators may say they are using *Science for All Americans* or the *Standards* as the basis for a state framework or a local program, but they commonly omit some aspects of the documents, for example technology, mathematics, or history.

Some dimensions of scientific literacy emerged in the last two chapters that clarify and elaborate scientific literacy. The use of scientific literacy implies a general education as opposed to specific education orientation for science programs. Although some science educators have written about science and general education, as a community, science educators have not understood, internalized, and developed this idea.

Scientific (and technological) literacy is best defined as a continuum of understanding about the natural and the designed world, from nominal to functional, conceptual and procedural, and multidimensional. This unique perspective broadens the concept to accommodate all students and gives direction to those responsible for curriculum, assessment, research, professional development, and teaching science to a broad range of students.

5 Establishing National Standards

By the mid-1990s, two major policy documents, the *National Science Education Standards* (National Research Council) released in 1995 and *Benchmarks for Science Literacy* (American Association for the Advancement of Science) released in 1993, addressed the reform issues. At the same time, state frameworks for science education were undergoing revision (Council of Chief State School Officers, 1995), and many local school districts were also developing and implementing standards. In March 1996, state governors and the chief executive officers of major companies held a second education summit. As a result of this gathering, they proclaimed continued support for high standards and agreed that the states should develop and implement standards for core subjects.

National Standards

Establishing national education goals and developing national standards for achieving those goals became an important strategy in the national effort to support educational reform. In 1989, state governors, chaired by Governor Terry E. Branstad of Iowa, endorsed national education goals at the annual meeting of the National Governors Association. At that gathering, then Governor Bill Clinton of Arkansas and his staff played a significant role in drafting the statements that would become *Goals 2000*. President George Bush expressed his support by forming the National Education Goals Panel, which soon gave rise to the National Council on Education Standards and Testing (NCEST). In 1992, the U.S. Department of Education initiated efforts to develop standards for several school subjects, including science, and began the process of translating broad purposes into more concrete policies to guide educational reform.

Mathematics educators and mathematicians introduced their national standards with the publication of *Curriculum and Evaluation Standards for School Mathematics* (National Council of Teachers of Mathematics, 1989) and *Everybody Counts:*

A Report to the Nation on the Future of Mathematics Education (National Research Council, 1989). These documents injected the word *standards* into public discussions of educational reform and contributed significantly to public understanding. They also added to the political turmoil over education standards.

Politicians have continued to express interest in improving American education. The National Education Goals Panel established the need for standards in different subjects and for performance-based assessments. When NCEST reported on the merit and feasibility of national standards and assessments, the National Council of Teachers of Mathematics (NCTM) Standards provided the proof the council needed. NCEST defined high expectations, not minimal competencies; they outlined a focus and a direction, not a national curriculum; they were national, not federal, voluntary, not mandatory, and dynamic, not static (National Council on Education Standards and Testing, 1992). As the political climate has changed, however, the emphasis on standards has changed, shifting primarily from the national level to the states. Attempts to eliminate national standards have not been successful. In fact, the release of the *Third International Mathematics and Science Study* (TIMSS) in late 1996 provided new support for standards based improvement of education (National Council on Education Standards and Testing, 1996; National Research Council, 1996; U.S. National Research Center for Third International Mathematics and Science Study, 1996).

Prelude to the National Science Education Standards *Project*

About the same time that *A Nation at Risk* was making a case for educational reform the NSB (1983) called for improvement in the scientific literacy of all citizens. In the 1980s, the AAAS initiated *Project 2061,* an ambitious multiyear undertaking that defined scientific literacy for high school graduates in *Science for All Americans.*

As the 1990s began, leaders in education, who recognized the need to coordinate reform efforts, introduced the idea of a nationwide systemic reform. At the NSF, Luther Williams, the assistant director for human resources, initiated State Systemic Reform programs. The NSF subsequently introduced Urban Systemic Initiatives, Local Systemic Initiatives, and Rural Systemic Initiatives, but the pressure to find a way to coordinate reform efforts increased with national concern about education. Policies initiated by President Bush have received continuing support under President Clinton and the leadership of Governor Roy Romer of Colorado and the National Governors Association.

In the spring of 1991, Dr. Bonnie Brunkhorst, then president of the NSTA, and William G. Aldridge, then executive director of the NSTA, backed by a unanimous vote of the NSTA Board, wrote to Dr. Frank Press, chair of the NRC, asking the NRC to coordinate the development of national science education standards. The presidents of several leading science and science education associations, the U.S. Secretary of Education, the assistant director for education and human resources at NSF, and the co-chairs of the National Education Goals Panel all encouraged the NRC to assume a leadership role. The NRC leadership proceeded with caution, recognizing that this effort represented a direct entry into the domain of science education and realizing the complexity of the undertaking. The subject matter of school science is drawn from a number of scientific and technical disciplines including biology, chemistry, physics, geology, astronomy, and engineering. Furthermore, science in K–12 education is not well established and its inclusion varies at elementary, middle, and high school levels. As the NRC leadership recognized, the multiplicity of science disciplines and the uneven presence of science in schools would make the development of national science education standards more complicated than the development of the mathematics standards. After extensive discussion with representatives of interested constituencies, the NRC undertook the task of developing and achieving a consensus on national science education standards.

The Project

On 16 September 1991, Lamar Alexander, the secretary of education, announced an award supporting NRC's project for the *National Science Education Standards.* Dr. James Ebert, then vice president of the National Academy of Sciences, was designated as chair of a National Committee on Science Education Standards and Assessment (NCSESA), the oversight body for the project.

Early in 1992, the NRC formed a Chair's Advisory Committee to assist in planning and directing the project. The committee consisted of representatives from the NSTA, the AAAS, NSRC, American Association of Physics Teachers (AAPT), American Chemical Society (ACS), Council of State Science Supervisors (CSSS), Earth Science Education Coalition (ESEC), and the National Association of Biology Teachers (NABT). This group participated directly in the process of identifying and recruiting codirectors of the project, chairs and volunteers for the working groups charged with drafting content, teaching, and assessment standards, and the national oversight committee (NCSESA).

Preparations for work on the standards began with the production of *Science Framework Summaries* (National Research Council, 1992), which was based on the work of the AAAS *Project 2061,* the NSTA *Scope, Sequence and Coordination of Secondary School Science,* and other projects, as well as state science frameworks, and science education standards from other countries. This compendium was made available to the national committee and the working groups.

The NCSESA first met in May 1992. The three working groups held intensive meetings over the summer of 1992. The content working group and the chairs of the three working groups devoted four weeks to constructing the initial framework for content standards. The working groups for assessment and teaching, which met later that summer, identified a variety of issues related to the development of standards in their respective domains.

By fall, the project leadership decided to develop an integrated document containing content, teaching, and assessment standards all displayed in mutually reinforcing ways. The working group chairs committed themselves to functioning as a team through the project. Another decision defined the critique and consensus to include updates on the project and draft materials for an intensive critique by science teachers, scientists, and others interested in science education. Using feedback from working papers released in October and November of 1992 and February and July of 1993, the project produced an initial report on content, teaching and assessment standards, and program and system standards.

Between May 1992 and spring 1994, several changes occurred in the project leadership. For health reasons, Dr. James Ebert stepped down as chair of NCSESA. Dr. Richard D. Klausner, a member of the National Academy of Sciences and chief of the Cell Biology and Metabolism branch of the National Institute of Child Health and Human Development at the National Institutes of Health (NIH) replaced him. In 1995, Dr. Klausner became director of the National Cancer Institute. Dr. Angelo Collins, a professor of science education at Florida State University, became project director.

In spring 1994, NRC printed the first complete draft of the *Standards*. This draft was the subject of extensive review by a number of groups, including NCSESA, focus groups convened by the Chairs' Advisory Committee, other selected groups and individuals, and most important, special panels convened by the NRC for the report review process. Approximately one thousand individuals participated in the initial review. The special NRC panels constituted the NRC report review and abided by the procedures approved by the Report Review Committee. Specific panels focused on content, teaching, assessment, program, and system standards. In addition, NRC convened a panel consisting exclusively of science teachers. This was a period of thorough, insightful, and critical feedback. The feedback included errors of technical details, omission of science disciplines and concepts, criticisms of the domains for standards, science topics that needed greater emphasis, the length and emphasis of the document, and "politically incorrect" statements.

After this period of review, the development team, which consisted of Audrey Champagne (assessment), Karen Worth (teaching), Rodger Bybee (content), and NRC staff Angelo Collins and Harold Pratt, had the task of collating all criticisms and determining exactly how they would be addressed. A document consisting of the criticisms and recommended changes was submitted to a special panel, termed the Executive Editorial Committee (EEC), which included representatives from NCSESA, CAC, the working groups for content, teaching, and assessment, and other representatives of NRC. This group approved the changes in June 1994.

Once the EEC had approved the proposed changes, the development team undertook a major revision of the national standards document. Members worked hard through the summer and fall of 1994 and released a draft for national review on 1 December 1994. Between December 1994 and February 1995 approximately forty thousand documents were distributed for review. In late March and early April 1995, the development team convened to analyze the results of the national review and make final recommendations. The *National Science Education Standards* was released in December 1995 (Collins, 1995).

Before proceeding further, I would like to make several points based on my work on the *National Science Education Standards.* I believe that the document has an essential role in the improvement of science education. In spite of political changes, the document will stand the test of time and prove to be an important organizer for states, districts, schools, and individual school personnel. I base these assertions on the process used to develop the *Standards.* First, the document was the outcome of the combined efforts of scientists, science educators, and science teachers. Many of the debates, especially in the development groups, centered on the appropriate accommodation of recommendations from specific perspectives, such as those of scientists or science teachers. In the end, I think we achieved the best balance possible. Second, I cannot overstate the importance of the review process. Even the early reports of 1992 and 1993, which many remember because of errors and omissions, served to provide feedback. Staying with the process contributed in substantial ways to a final document that has received little criticism. Finally, interpreting and implementing the document require enlightened professionals, and I am convinced that this will be achieved through study, review, and thought about what standards are (and are not), why they are important, how they should be used, and what they will provide the science education community.

An Overview

The *National Science Education Standards* includes standards for teaching, professional development, assessment, content, program, and system, all directed toward the goal of improving science education and encouraging students to achieve higher levels of scientific literacy. By developing standards in these six domains, the document accommodates the systemic nature of educational reform. Those of us who worked on the project recognized the essential role of science teachers, so it should be no surprise to see standards addressing science teaching. The Teaching Standards provide a vision of what science teachers need to understand and what they need to do to assure adequate learning experiences for all students:

- Teachers of science plan an inquiry-based science program for their students.

- Teachers of science guide and facilitate learning.

- Teachers of science engage in ongoing assessment of their teaching and of student learning.

- Teachers of science design and manage learning environments that provide students with the time, space, and resources needed for learning science.

- Teachers of science develop communities of science learners that reflect the intellectual rigor of scientific inquiry and the attitudes and social values conducive to science learning.

- Teachers of science actively participate in the ongoing planning and development of the school science program.

The Standards for Professional Development focus on the continuing growth of the science teacher's knowledge and teaching skills:

- Professional development requires learning science content through the perspectives and methods of inquiry.

- Professional development requires integrating knowledge of science, learning, and pedagogy and applying that understanding to science teaching.

- Professional development requires building the understanding and skills that allow students to engage in lifelong learning.

- Professional development programs must be coherent and integrated.

The Assessment Standards provide criteria for judging the quality and fairness of assessment procedures:

- Assessments are consistent with the decisions they are intended to inform.

- The technical quality of the data collected is well matched to the consequences of the decisions and actions taken on the basis of its interpretation.

- Assessment practices are fair.

- Assessment includes both student achievement and opportunity to learn science.

- The inferences made from assessments of student achievement and opportunity to learn are sound.

The Assessment Standards recommend consistency with the Content Standards and broaden the view of assessment in three areas. First, they expand the definition of assessment processes to include those in schools, in the states, and in the nation. Second, they support the science teacher's responsibility for planning and interpreting assessments at different levels, including classrooms, districts, states, and national organizations. Finally, they direct attention to the students' achievement of scientific literacy, the teachers' opportunities to teach, and the schools' provision of opportunities for students to learn science content.

The Content Standards define what all students should know and be able to do. They describe the knowledge, fundamental understandings, and abilities that students should develop during their thirteen years of formal education. Science content includes the following categories:

unifying concepts and processes

science as inquiry

physical science

life science

earth and space science

science and technology

science in personal and social perspective

history and nature of science

The Content Standards are discussed in greater detail in Chapter 6, but two points are worth mentioning here: first, the Content Standards define scientific literacy, but they do not define how the content should be organized into a science curriculum.

In developing Content, Teaching, and Assessment Standards, the various working groups always recognized the interdependence among standards. How students work together in immediate learning environments, such as the classroom, the school, and the community, defines their opportunity to learn. How teachers are supported by policies beyond the immediate learning environment, for instance, in the district, the state, and the nation, informs their opportunity to teach. Fairly early in the project, the leadership also recognized the need to include Program and System Standards. The overarching goals of Program and System Standards are first, the coordination of content, teaching, and assessment with resources and second, the alignment of teaching and assessment with the goals of science education.

Program Standards outline procedures for designing, and criteria for judging, the quality of school science programs. These standards address issues that affect science teachers and students directly: the quantity and quality of opportunities to achieve the Content Standards; connections between science and mathematics, specifically the NCTM Standards; the allocation of resources; the consistency of the program with content, teaching, and assessment standards; and the coherence of the science program across the K–12 continuum. The principles underlying Program Standards are as follows:

- All elements are consistent with the other *National Science Education Standards* and across grade levels.

- The curriculum in science for all students in grades K–12 includes all content standards in a variety of curricula that are developmentally appropriate, interesting, relevant to students' lives, organized around inquiry, and connected with other school subjects.

- The science program is coordinated with mathematics education.

- The science program provides appropriate and sufficient resources to all students.

- The science program provides equitable opportunities for all students to learn the *National Science Education Standards*.

- Schools are communities that encourage, support, and sustain teachers.

Science education is one part of a larger education system that interacts with economic, social, and political systems. Science classrooms are affected by the economic, social, and political conditions in their local communities. The nation's scientific enterprise is a unique aspect of science education, just as science education is an important contributor to the scientific enterprise. These systems depend on the rational allocation of resources and on coordinated effort.

The System Standards are intended to guide and coordinate the educational system toward the goal of a scientifically literate citizenry. They provide guidance for policies, programs, and actions outside the immediate learning environment that will produce high-quality science programs with congruent content, teaching, and assessment. Recognizing the vital role of science teachers, the System Standards acknowledge the fact that various aspects of the educational system must support their efforts in the classroom. The national, state, and local systems will have to assume responsibility for providing financial, material, and intellectual resources to achieve the goal of scientific literacy for all students. The organizing principles are as follows:

- Policies that influence science education must be consistent with teaching, professional development, assessment, content, and program standards.

- Policies should be coordinated within and across agencies, institutions, and organizations.

- Policies need to be sustained over sufficient time.

- Policies must be supported with resources.

- Policies must be equitable.

- All policy instruments must be reviewed for possible unintended effects on science education.

- Responsible individuals take the opportunity to achieve the new vision of science education portrayed in the standards.

Science teachers, supervisors, curriculum developers, administrators, scientists, and parents can refer to the national standards to improve science education and scientific literacy. They should recognize, however, that the Content Standards do not prescribe curriculum; the Assessment Standards are not examinations; and the Teaching Standards are not licensure requirements. Rather, these standards present a vision of the changes and improvements their application would make possible.

Establishing the *National Science Education Standards* identified issues that represent profound challenges for the science education community. These challenges include the following:

- The goal that all students will achieve all science content is different from current programs and practices.

- Science content in the national standards is different from a science curriculum.

- Learning science content is different from teaching science lessons.

- Understanding science content is different from knowing scientific information.

Science education must be inclusive of all students, regardless of gender, race, culture, physical or learning disabilities, future aspirations, or interest in science. And standards imply that all students should develop understanding and skill in all content areas. These principles of inclusion and attainment have direct implications for curriculum design, allocation of resources, teaching strategies, and assessment practices. I have noted the science education community emphasis on "science for all" in the 1990s. This slogan takes on a different spin when the implied goal is that all students should achieve the fundamental understandings and abilities described in the Content Standards. The degree to which we achieve this vision will be a significant measure of the success of this reform effort.

One of the most significant misconceptions about the Content Standards equates content categories with the scope, sequence, and emphasis of the science curriculum. A science curriculum includes a content emphasis, such as scientific inquiry and reasoning abilities or STS. But a science curriculum also includes instructional approaches and assessment strategies. Translating the Content Standards to a curriculum thus requires a whole series of decisions by curriculum developers, school districts, and science teachers.

Learning science content is not the same as teaching a science lesson. The national standards identify what students should know and be able to do at grades 4, 8, and 12, and emphasize science concepts and skills. This emphasis on students learning contrasts markedly with an emphasis on experiencing lessons with similar titles. The issue is not whether science teachers incorporate a lesson on the "Origin and Evolution of the Earth System," even if it is "hands-on, minds-on." The question is, Did students develop understandings identified with the "Origin and Evolution of the Earth System"?

All educators need to recognize the difference between teaching information and developing students' understanding of concepts. For some time, school science programs, textbooks, and teachers have focused on knowing information and facts. This information is essential, but what the Content Standards do is to challenge science teachers to help students develop understanding. A student might be able to list the parts of a cell but not understand the relationship of cells to the organism's function. Or a student might be able to describe the parts of atoms but not understand the conservation of energy. My intention here is not to engage in a debate about words but to highlight the need to focus science teaching on higher levels of thinking, richer conceptual meaning, and important applications of knowledge.

Learning the science described by the standards means understanding science content and developing the basic skills of inquiry and design. Science teaching will move beyond informational lessons and encourage students to apply their understanding to situations in which they have to think critically, analyze logically, and reason scientifically. "Science as inquiry," for example, requires that students find and use evidence, formulate explanations, compare their explanations with current scientific knowledge, and apply their explanations to unique situations.

If we are serious about improving science education we will have to recognize the deep changes in curriculum, instruction, and assessment, and, by extension, the significant changes in school science programs and in the educational system the national standards imply. The fundamental characteristics of the *National Science Education Standards* can be summarized in the following points:

The Standards

- Define the understanding of science that all students, without regard to background, future aspirations, or prior interest in science, should develop.

- Present criteria for judging science education content and programs at the K–4, 5–8, and 9–12 levels, including learning goals, design features, instructional approaches, and assessment characteristics.

- Include all natural sciences and their interrelationships, as well as the natural science connections with technology, science- and technology-related social challenges, and the history and nature of science.

- Incorporate recommendations for the preparation and continuing professional development of teachers, including resources needed to enable teachers to meet the learning goals.

- Propose a long-term vision for science education, some elements of which can be incorporated almost immediately in most places, others of which will require substantial changes in the structure, roles, organization, and context of school learning before they can be implemented.

- Provide criteria for judging models, benchmarks, frameworks, curricula, and learning experiences developed under the guidelines of ongoing national projects, or state frameworks, or local district-, school-, or teacher-designed initiatives.

- Provide criteria for judging teaching and the opportunities to learn valued science (including such resources as instructional materials, educational technologies, and assessment methods) and science education programs at all levels.

Benchmarks and Standards

In the late 1980s and early 1990s, before the *Standards,* several policy statements and curriculum frameworks significantly influenced state and local reform of school science programs. Those frameworks include the report *Science for All Americans* (American Association for the Advancement of Science, 1989) and the subsequent publication of *Benchmarks for Science Literacy* in 1993; NSTA's (1989) project *Scope, Sequence, and Coordination;* the National Center for Improving Science Education (NCISE) reports on elementary and middle-level education (Bybee et al., 1989a, 1990) and secondary education (Champagne et al., 1991). The AAAS frameworks have had a considerable impact on the reform of science education. They have also generated persistent questions because they coexist with the *Standards*.

Project 2061

In the 1980s, F. James Rutherford, the chief education officer for the AAAS, established *Project 2061* in order to take a long-term, large-scale view of educational reform in the sciences in relation to the goal of scientific literacy. The core of *Science for All Americans* consists of the recommendations of a distinguished group of scientists and educators about what understandings and habits of mind are essential for all citizens in a scientifically literate society. Scientific literacy, which embraces science, mathematics, and technology, is a central goal of science education, yet general scientific literacy eludes American society. In preparing its recommendations, *Project 2061* staff drew on the reports of five independent study panels and sought the advice of a large and diverse array of consultants and reviewers, including scientists, engineers, mathematicians, historians, and educators. The process took more than three years, involved hundreds of individuals, and culminated in *Science for All Americans.* Its recommendations, therefore, represent basic learning goals for American students. One premise of *Project 2061* is that the schools do not need to teach more content, they need to teach less content *better. Science for All Americans* outlined the basic dimensions of scientific literacy:

- being familiar with the natural world and recognizing its diversity and its unity

- understanding the concepts and principles of science

- being aware of some of the ways in which science, mathematics, and technology are interdependent

- knowing that science, mathematics, and technology are human enterprises that reflect human strengths and limitations

- developing a capacity for scientific ways of thinking

- using scientific knowledge and ways of thinking for individual and social purposes

Science for All Americans covers an array of topics, many of which are already common in school curricula (for example, the structure of matter, the basic functions of cells, the prevention of disease, communications technology, and different uses of numbers). But its treatment of such topics differs from traditional approaches in two ways. First, boundaries between traditional subject-matter categories are softened and connections are emphasized through the use of important conceptual themes, such as systems, evolution, cycles, and energy. Transformations of energy, for example, occur in physical, biological, and technological systems, and evolutionary change occurs in stars, populations of organisms, and societies. Second, the amount of detail students are expected to learn is less than in traditional science, mathematics, and technology courses because this approach emphasizes key concepts and thinking skills instead of specialized vocabulary and rote procedures. *Science for All Americans* not only makes sense at a simple level, it also provides a lasting foundation for further learning. Details are used to enhance, not guarantee, students' understanding of a general idea.

Science for All Americans also includes topics uncommon in school curricula, among them the nature of the scientific enterprise and how science, mathematics, and technology relate to one another and to the social system in general. Even the title conveys an important symbolic value. This volume established a new "worldview" for science education: accommodating the needs and aspirations of all students. Although to some, the phrase "science for all" is a slogan, I would claim that it has power to reorient school science language and thinking toward general education purposes.

In 1993, *Project 2061* released *Benchmarks for Science Literacy,* which was based on *Science for All Americans.* The benchmarks comprise the specific outcomes of science education. Many local school districts and state agencies, and some national organizations have begun using them for frameworks, syllabi, and curriculum models. I will elaborate on some of the details of *Benchmarks for Science Literacy* because it shares similarities with the *National Science Education Standards,* a situation that stimulates questions but also causes some confusion. The chapters in *Benchmarks for Science Literacy* correspond to those in *Science for All Americans* as, for the most part, do the sections within the chapters. The chapter titles suggest its range:

Benchmarks

1. The Nature of Science

2. The Nature of Mathematics

3. The Nature of Technology

4. The Physical Setting

5. The Living Environment

6. The Human Organism

7. Human Society

8. The Designed World

9. The Mathematical World

10. Historical Perspectives

Each chapter begins with a quotation from *Science for All Americans* and then proceeds to a general discussion of the ideas students should learn and the types of experiences that might facilitate that learning.

At the section level, for example, Cells in Chapter 5, "The Living Environment," there is a discussion of the common misconceptions and the difficulties students may have in learning about cells. There is also some discussion and clarification of the idea and some development of the idea at grade levels K–2, 3–5, 6–8, and 9–12. For example, the benchmarks for "Interdependence of Life" at grades 3–5 are as follows:

By the end of the fifth grade, students should know that

- For any particular environment, some kinds of plants and animals survive well, some survive less well, and some cannot survive at all.

- Insects and various other organisms depend on dead plant and animal material for food.

- Organisms interact with one another in various ways besides providing food. Many plants depend on animals for carrying their pollen to other plants or for dispersing their seeds.

- Changes in an organism's habitat are sometimes beneficial to it and sometimes harmful.

- Most microorganisms do not cause disease, and many are beneficial.

The benchmark statements contain suggestions as well as an orientation for curriculum and instruction, and the discussion provides indicators of experience over the grade span. The actual benchmarks are phrased as outcomes: "By the end of the fifth grade, students should know that . . ." and the language is generally nontechnical, while indicating a sophistication appropriate for the grade level.

In light of the framework for scientific literacy (see Chapter 4), a notable characteristic of the benchmarks is that they use technical words: *environment, survival, plants, animals, insects, pollen, seeds, habitat, microorganisms,* and *disease.* I would extend this list by adding several words that, strictly speaking, may not be considered technical vocabulary, but which require specific understanding in the context:

interact, depend, beneficial, harmful, and *cause.* I point this out to show the appropriate role of vocabulary in scientific literacy, although the primary emphasis in *Benchmarks* is not vocabulary.

Benchmarks for Science Literacy provides curriculum developers with a unique tool for designing and developing school science programs. It

- serves as a resource for designing a curriculum that meets the criteria for science literacy described in *Science for All Americans*

- facilitates curriculum diversity because it provides a compendium of science literacy goals that can be organized in unique ways by curriculum designers

- establishes a set of outcomes for threshold levels of understanding and ability that all students are expected to reach at their designated grade levels

- identifies a common core of learning that contributes to the science literacy of all students

- uses language that indicates the level, nature, and sophistication of student learning

- includes suggestions for instruction to clarify the meaning and intent of the benchmarks

- emphasizes educational research and the recommendations of classroom teachers

Relationships Between the *Standards* and *Benchmarks*

As soon as the work on the *National Science Education Standards* began, individuals asked about the relationship between the *Standards* and *Benchmarks.* These questions quickly extended to the relationship of these national projects to state and local standards and frameworks. Other projects, primarily NSTA's *Scope, Sequence, and Coordination,* are essentially curriculum projects or programs that represent the implementation of standards. The *Standards* and *Benchmarks* are policy statements.

Similarities Between the Standards and Project 2061

Similarities indicate what is considered fundamental in the reform effort. *Project 2061* is a long-term reform initiative to transform K–12 education through a coordinated set of reform tools, including the goals of *Science for All Americans,* the policies of *Benchmarks for Science Literacy,* and blueprints—for assessment, equity, curriculum, higher education, school organization, and finance. The various tools of *Project 2061* are for school districts to use in developing their own curricula so that all students achieve science literacy. *Project 2061* began in 1985. *National Science*

Education Standards is a short-term reform initiative designed to produce long-term improvements in science education. The *National Science Education Standards* contain a coordinated set of policies for content, teaching, assessment, programs, and the educational system. The national standards will guide a transformation of the science education system in productive and socially responsible ways. The national standards leave the responsibilities for specific curriculum materials and assessment strategies to states and local school districts. Achieving higher levels of science literacy is the overarching goal of the standards. The *Standards* project began in 1991.

In large measure, similarities are not surprising, since each project was developed with recognition, information, and advice from the others. (The national standards project drew from *Project 2061* materials, for example, while Jim Rutherford served on the Chairs' Advisory Committee for the *Standards* project.) Both projects were nationally oriented, multiyear initiatives to improve science education. They

- focused on the goal of achieving higher levels of science literacy for all students

- promoted the successful involvement of traditionally underrepresented groups that have been alienated by science education programs

- advocated general education goals that integrate personal and social perspectives with science concepts and processes

- recognized contemporary approaches to teaching and learning

- left the actual design and implementation of curriculum, instruction, and assessment to states and local school districts

- incorporated technology, history, and the nature of science as components of science education

- emphasized a deeper understanding of science concepts and procedures rather than a broader coverage for science topics

- involved broad-based review by science teachers, science educators, scientists, and communities, during development of specific products and materials

- recognized the important role of science teachers and their need for continual professional development and support

Differences Between the Standards and Project 2061

Despite their common elements, such as their general agreement on what constitutes scientific literacy, how to enhance learning, and who is responsible for reform, questions about their differences persist, however insignificant these differences are: Why are there two (or three) projects? Which project(s) should we use in our state or school district? Table 5.1 offers a useful comparison.

TABLE 5.1 *Major Differences Between* Project 2061 *and* Standards

Project 2061	*National Science Education Standards*
Involves systemic reform in K–12 education for science literacy.	Involves content, teaching, and assessment, program, and system standards for systemic reform of K–12 science education.
Addresses the reform in terms of the content specified by *Science for All Americans* and the outcomes in *Benchmarks for Science Literacy.* Sequencing and the spacing of content in a curriculum are considerations, but not definitions.	Addresses reform through standards for judging programs that incorporate learning goals, content, design features, assessment characteristics, and support for teachers, including resources needed to meet the learning goals.
Focuses on K–12 curricula, with specific benchmarks for science literacy at grades 2, 5, 8, and 12 that identify and sequence outcomes and their content interconnections.	Focuses on K–4, 5–8, and 9–12 grade spans, with recognition of undergraduate, graduate, and continuing professional development of science teachers.
Incorporates the interdependence of science disciplines and integrates through connections among the science disciplines and with other disciplines, such as the arts and the humanities.	Incorporates all natural sciences and their relationship to technology, society, history, and other school subjects such as mathematics.
Recognizes the need for long-term guidance of the entire K–12 system through blueprints on teacher education, assessment, policy, and other school issues.	Addresses the interdependence of the science disciplines and connections with technology, social science, mathematics, and history.
	Recognizes the need for a long-term vision and a short-term solution for the transformation of science education.
Provides different models for restructuring curriculum: in parallel arrangement with little overlap among subjects; integrated around issues or phenomena; or in a mosaic, bound by a variety of organizing principles.	Provides criteria for assessing curricula developed by curriculum-development groups, state frameworks, local district, school- or teacher-designed programs.
Designs a coordinated set of reform tools, the basic components of curricula with alternative approaches to teaching and assessment, for school districts to use in developing their own curricula.	Design criteria for judging opportunities to learn, assessments of students' achievement of science and teachers' teaching (including such resources as instructional materials and assessment methods) and science education programs at all levels.

A Perspective on the Standards and Project 2061

Focusing on the actual reports from these two projects clarifies one of the most significant differences between them: the *Standards* provides a comprehensive set of criteria, principles, and specifications for judging current science curriculum, teaching, and assessment, for aligning current programs with standards, and for developing new programs and practices to achieve higher levels of scientific literacy. The document contains all the standards for teaching, assessment, content, program, and system.

The *Standards* and *Benchmarks* are comparable sets of policies. Both provide similar statements of *content outcomes*. An analysis completed by *Project 2061* concluded that there was a "consensus on content" (American Association for the Advancement of Science, 1995). The statements of content—the actual standards and benchmark compare on "grain size," length, and comprehensive treatment of the disciplines. Overall, I would estimate that more than 90 percent of the content associated with the traditional disciplines of physical, life, and earth sciences overlaps. On the surface there may seem to be differences, but closer analysis shows that the documents are quite comparable. The overlap is easily explained. First, the *Standards* content working group and development team had access to the preliminary drafts and to the published benchmarks. Second, the group made a deliberate effort to align standards and benchmarks while revising over the summer and fall of 1994 by maintaining the integrity and the unique framework of the respective documents while aligning the standards statements, particularly the fundamental understandings. The introduction to the *National Science Education Standards* (National Research Council, 1996) acknowledges the congruence of standards and benchmarks.

> The many individuals who developed the content standards sections of the *National Science Education Standards* have made independent use and interpretation of the statements of what all students should know and be able to do that are published in *Science for All Americans* and *Benchmarks for Science Literacy*. The National Research Council of the National Academy of Sciences gratefully acknowledges its indebtedness to the seminal work by the American Association for the Advancement of Science's *Project 2061* and believes that use of *Benchmarks for Science Literacy* by state framework committees, school and school-district curriculum committees, and developers of instructional and assessment materials complies fully with the spirit of the content standards. (p. 15)

The extensive reviews of both carried out by scientists, science educators, and science teachers contributed substantially to the convergence of the two documents. Feedback, in particular the negative feedback, served to focus the categories, topics, concepts, developmental appropriateness, and the concreteness or specificity of content statements. (*Benchmarks*, for example, contains a chapter on issues and language that applies to that volume but could well have been written for *Standards*.)

For individuals developing the *Standards* and, I would imagine, the *Benchmarks* there was a constant tension between the need to develop a content framework that fully and adequately represented science and technology (but completely and appropriately avoided defining a curriculum) and requests to be concrete, specific, and detailed about content. Broad concept statements would not do, yet specific statements began to read like behavioral objectives, classroom experiences, and teaching activities. This tension between what I will characterize as the conceptual and the practical focused efforts on policies to guide reform. I use the term *constructive ambiguity* to characterize statements of content, teaching, and assessment in the *Standards*. These statements necessarily had to be concrete and clear enough to provide constructive direction for those who would use these reports. At the same time, they had to remain slightly ambiguous because the individuals who would actually refer to the *Standards* and *Benchmarks* would need to be free to design curriculum frameworks and materials.

One of the most significant differences between the *Standards* and *Benchmarks* is the result of comprehensiveness; the *Standards* included standards for teaching, professional development, assessment, content, programs, and the educational system on publication in 1996. *Project 2061* has future plans for a comprehensive set of blueprints and designs. The *Standards* address three aspects of science inquiry, the ability to carry out a scientific inquiry and the understanding gained from inquiry. The third aspect of inquiry has to do with teaching strategies. All those who worked on the *Standards* believed in inquiry as a unique aspect of science and an important method of teaching science. Our work on the inquiry standards appealed to the early work of Joseph Schwab and others (Schwab, 1966) and applied the contemporary thinking of Richard Duschl (1990) and others (Kyle, 1980; Duschl and Hamilton, 1992; Costenson and Lawson, 1986). The *Standards* emphasized that students should actually conduct their own inquiries as a part of their school science experience. To be clear, *Project 2061* does not oppose inquiry, although they have criticized the *Standards* for the time and the effort devoted to inquiry. Their emphasis is on "Habits of Mind" and developing an understanding of scientific inquiry.

I would also say that the *Standards* gives more emphasis to science in personal and social perspectives. *Benchmarks* contains some of the same issues but with less emphasis.

In 1994 and 1995, the Council of Chief State School Officers and SRI International conducted an evaluation of state curriculum frameworks in science and mathematics. I participated in the study as a panel member reviewing curriculum frameworks for science. Four primary questions guided the study: What is the status of curriculum frameworks in all fifty states? What curriculum content is recommended at the state level in mathematics and science frameworks for specific grades? What are the purpose and structure of state frameworks? How are state frameworks developed and implemented?

The study of state curriculum frameworks gives considerable support to the fact that states are using the *Standards* and *Benchmarks* in the revision of frameworks.

Because the *Benchmarks* has been available since 1993, more states have referred to it as they developed their framework or syllabus. Vision statements and conceptual organizers for science content are generally consistent with national standards. Technology and the history and nature of science are the areas that tend to be omitted or underemphasized. As expected, state frameworks have gone a step beyond national standards and recommended a curriculum orientation, specific topics, pedagogy, and in many cases, assessments (Council of Chief State School Officers, 1995).

Some are upset by the *Standards* and *Benchmarks,* but I consider it a good thing to have both. Debate, review, and even conflict are healthy for the profession, and *Standards* and *Benchmarks* provide the materials for critical thinking and careful analysis by the science education community. Most individuals who are critical of the *fact* of the documents—as opposed to the documents themselves—seem upset that they have to review the documents and identify the similarities and differences. In some cases, it is clear that they have not even reviewed them. The consensus on content far exceeds any differences and presents the thinking of thousands of individuals and the reviews and critiques of tens of thousands. The projects have done the hard work of sifting and sorting that is most fundamental and important for students to understand. Individuals can avail themselves of the opportunity to attend to and use the *Standards* and *Benchmarks* and assess the fidelity of their programs to these policy statements. Both support the incorporation of technology, the history of science, and the nature of science into school science programs. Yet state frameworks, school syllabi, and many contemporary curricula give little attention to these categories or omit them entirely.

Establishing Standards for Scientific Literacy

Many local school districts and states either have developed or are in the process of developing standards. Although the 1996 Education Summit shifted the locus of control to the states, it did not reduce the essential place of standards. The national standards will also continue to influence this system. This is especially supported by the average and below average scores of U.S. eighth-grade students on science and mathematics achievement on TIMSS (National Center for Educational Statistics, 1996).

The Power, Place, and Promise of Standards

Establishing standards for scientific literacy offers power and promise: power in the shared vision that standards convey and promise in their potential to coordinate and regulate the reform of science education. For too long, we have gone about the business of changing science programs and practices without a clear, much less a shared, vision of what we were trying to accomplish. In the early 1990s, the science community began taking action toward the goal of achieving scientific literacy, and the need for change has gained acceptance. Standards have given the community an active and productive set of goals at national, state, and local levels. An added benefit has been the large-scale acceptance of the idea of standards and their eventual use in school science programs and classroom practices.

Once established, standards serve to guide reform through the coordination of state and district policies, programs, and practices. This coordination occurs at the national level, and it can also be seen in the priorities of agencies, such as the NSF, the National Aeronautics and Space Administration (NASA), the NIH, and the U.S. Department of Education.

The March 1996 Education Summit and release of TIMSS late in the same year clearly established the need and place of standards in contemporary reform. The policy statement issued at the national summit and the briefing book provided for the governors and CEOs addressed such issues as explicit expectations and school accountability, and why standards can improve student achievement. The statement also incorporated a commitment to set clear academic standards in core subjects. Reflecting a widespread sentiment, the governors and CEOs made educationally naive assertions that national standards are dead. For a nation worried about its international competitiveness and concerned about its uncoordinated and incoherent educational system, the thought of separate standards for fifty states and sixteen thousand school districts seems incomprehensible. The governors want to improve student achievement, so they call for high standards. But what counts as *high?* More vocabulary? Fundamental concepts? The potential for even greater inconsistency increased with the accelerated development of standards during the two-year period 1996–98. There could be inconsistencies, for example, in the core content of science. Will all states include evolution? How about the laws of thermodynamics? Some states might give greater emphasis to *integrated science,* thus omitting some core concepts, while others might stress *discipline-based science,* thus omitting unifying conceptual themes.

It is worth noting that we already have a national science curriculum. It is comprised of extant textbooks. This is the curriculum that, one can reasonably assume, contributes to our very low levels of achievement and our poor showing in international comparisons (National Center for Educational Statistics, 1996). It has been "dumbed down" and thus, is highly criticized.

The governors have agreed that to "reallocate seems sufficient to support implementation of those standards within a clear timetable for full implementation" and that "such funds should be available for the essential professional development, infrastructure, and new technologies needed to meet these goals" (Lawton, 1996), and it is no stretch of the imagination to conclude that the national standards will figure prominently in state initiatives. The wisest thing the states could do is begin with the science and mathematics standards and direct all reallocated resources to professional development, in particular building the capacity for continuous professional development in school districts.

Why do we persist in using the example of the history and the English standards to conclude that we should not have national standards? It would be just as unreasonable to use the mathematics and science standards to conclude that we *should* have national standards. I would even cite the case of the history standards to show that the process of review, criticism, and revision works. The second edition of the history standards is, by most reports, acceptable and usable.

In my view, national standards are essential. Although the way curriculum and teaching are organized may show considerable variation, even if it currently varies little across the states, the core content of science need not vary. The basic concepts of physical, life, and earth science; the role of inquiry and design; understanding society; and the history and nature of science should be as common in New York and California as in North Dakota and Mississippi. After all, the states have agreed to some things that serve the nation's interest as a whole (highways, for example, have common boundaries at the borders of states).

Time will turn those with responsibility for developing state's standards toward the idea of national standards, and we will achieve what the nation agreed to in the first place—national standards, not a national curriculum. We will also have a more consistent and coherent system for achieving higher levels of scientific literacy. If the politicians persist in their views, we will have even more support for high levels of scientific literacy, a goal that includes logic and reasoning as well as an understanding of a few basic ideas.

Standards and the Commitment to All Students

Part of the power and promise of standards is their potential to fulfill the social commitment of educating all students. Standards for science education define the level of understanding that all students should develop. State and local standards should embody the belief that all students can learn science. Science education standards should be designed to encourage all students to study science throughout their school years and to pursue careers in science. Adopting the science for all goal will encourage the participation of all students in challenging learning opportunities. Standards should advocate the inclusion of those who traditionally received little encouragement or opportunity to learn science—women and girls, all racial and ethnic groups, the physically and educationally challenged, and those with limited proficiency in English—as well as the gifted and talented who have traditionally been achievers.

The Importance of Understanding Standards

Most people still cannot differentiate between voluntary national standards and mandatory federal laws. Similar misunderstandings about the development and use of standards flourish. State and local educators should be clear about the role of standards and the needs they meet, about what standards are and what they are not. Standards are policy guides; they are not curriculum materials or lesson plans that describe what students should know, value, and do as a result of their science education. Standards should outline the content domain of science, which generally includes concepts associated with the physical, life, and earth sciences. In addition, it is very important, and consistent with the goal of achieving scientific literacy, to include inquiry, technology, the history and nature of science, and science in personal and social perspectives. It also is a good idea to state the standards as outcomes—what students should know and do—and not as experiences they should have, because outcomes allow greater leeway for curriculum developers and classroom teachers.

I would address issues of teaching and assessment in tandem with the development of content standards. It is important to establish the fact that the opportunities for learning will have to be established and to define what will count as "meeting the standards." Performance standards at the state and local levels should be consistent with content standards.

What can standards do and not do? Standards are policies that should guide and coordinate science reform efforts. They should provide developmentally appropriate vocabulary, concepts, processes, and contexts for elementary, middle, and high schools. Standards should include guidelines for inclusion, for example, evolution should be included in school science programs and creationism should not. They should also enlarge the orientation of the science program by addressing teaching issues, approaches to learning, the use of technology, and assessment strategies. When implemented, standards should focus, direct, and guide, but not eliminate the responsibilities of those who develop curricula and teach science.

Finally, in the process of establishing standards, someone always asks, Haven't we done this before? The answer is yes and no. We have often redefined the content of the science curriculum: in the 1950s and 1960s, for example, we made a major attempt to update the science curriculum. But it does not necessarily follow that the entire experience of establishing standards is useless or that an earlier framework is preferable. For there is a second question: If we did it before, did it work? For this the answer is less clear, and may, in fact, be no. This suggests that achieving scientific literacy may involve more than simply updating content. The *Standards* emphasized teaching, assessment, school science program, and the educational system with the hope that this wider perspective would contribute to a substantive improvement in science education. Establishing standards for scientific literacy, incorporating teaching and assessment, and addressing the larger educational system have *not* been tried before.

Conclusion

Establishing national standards provides policies to guide the reform of science education. The policies are an important and intermediate step between the purpose of achieving scientific literacy and the programs and practices of schools and science teachers, respectively. As difficult and complex as the process of establishing standards was, it represents only one step in the long journey of reform. The standards and benchmarks provide a map of the territory; they are not the territory, and they only hint at the difficulties of travel toward the final destination.

6

Creating a Vision of Scientific Literacy

In 1992, the NCSESA challenged all those working on national standards to "create a vision for the scientifically literate person and standards for science education that, when established, would allow the vision to become reality. The standards, founded in exemplary practice and contemporary views of learning, science, society, and schooling, will serve to guide the science education system toward its goal of a scientifically literate citizenry in productive and socially responsible ways" (National Research Council, 1995). This challenge begins with a vision of the scientifically literate person. The national standards in general and the Content Standards in particular answer what I have termed the Sisyphean question in science education: What should the scientifically literate person know, value, and be able to do—as a citizen? What this also suggests is that the standards should be concrete enough to allow us to ascertain our degree of success in achieving that vision. Statements of science content must connect with assessment. The committee also draws a direct connection to science teaching and student learning that extends beyond content, teaching, and assessment to the science education system as a whole. To meet this mandate, the *National Science Education Standards* (National Research Council, 1996a) includes content, teaching, and assessment, but it also recognizes a larger context for improving science education through standards for the school science programs and the educational system.

An Achievable Challenge

Is this an appropriate challenge for science education? The answer is yes. Citizens recognized the need to improve education long before the Goals 2000: Educate America Act of 1994, yet this act underscored the vital place of science in general educational reform. Having stated the broad purpose—achieving scientific literacy—the task is to provide a more concrete statement of exactly what it is students should know, value, and do. The challenge extends further to programmatic and systemic issues and structural and functional elements.

Is this challenge achievable? The answer to this question must also be yes, but, it should be clear that the *Standards* represents only one small aspect of a process that involves the broad scope of educational systems. Responsibility for achieving scientific literacy extends far beyond science teachers and students to ordinary citizen taxpayers and local, state, and federal government and elected officials who must join with educators, scientists, and business leaders to make the standards a reality. Society's institutions and governmental organizations share the obligation to provide schools with the necessary fiscal resources so that teachers have the opportunity to teach science and all students have the opportunity to learn science. Along with fiscal support, society must provide intellectual and political support for science education, demonstrating the value it places on scientific literacy as well as its belief that all students can achieve reasonable levels of scientific literacy.

Science Content

The Content Standards (National Research Council, 1996a) define general scientific literacy: as the result of inquiry-oriented activities, students should develop an understanding of fundamental concepts and abilities in the natural sciences. The Content Standards also point out important connections between science and technology, personal and social perspectives, and the history and nature of science. They fall into eight categories:

- science as inquiry
- physical science
- life science
- earth and space science
- science and technology
- science in personal and social perspectives
- history and nature of science
- unifying concepts and processes

The eighth category, unifying concepts and processes, integrates topics and content across all grade levels K–12. The other seven include standards for grade levels K–4, 5–9, and 9–12, and these grade-level clusters are also consistent with those defined in national standards for other disciplines, such as mathematics, technology, and geography.

Science as Inquiry

Inquiry is the basis of scientific literacy, and thus an essential component of school science programs. For all three grade-level clusters, the standards highlight the ability to carry out and understand the process of scientific inquiry. As a result of their experience, students should learn to think and act in ways associated with the processes of inquiry: asking questions, planning and conducting an investigation using appropriate tools and techniques, thinking critically and logically about the relationships between evidence and explanations, constructing and analyzing alternative explanations, and communicating scientific arguments. In addition, students should understand that scientific inquiry is basic to the discipline. Early in the development of inquiry standards, Dr. Richard Duschl encouraged the study group to focus on students' ability to think and to use observations and knowledge to formulate scientific explanations. The national standards challenge educators to move beyond "science as a process," in which students learn skills (observing, inferring, and hypothesizing) and to combine these skills with scientific knowledge, scientific reasoning, and critical thinking to construct a richer understanding of science. The new vision gives greater emphasis to cognitive abilities and less to observing, inferring, classifying, controlling variables, and forming hypotheses, skills that in recent years have been taught as ends in themselves. The shift in emphasis does not eliminate them but, rather, regards them as a means to greater conceptual understanding. According to the *Standards,* engaging students in inquiry serves five essential functions:

- It assists the development of understanding of scientific concepts.

- It helps students "know how we know" in science.

- It develops an understanding of the nature of science.

- It develops skills necessary to become independent inquirers about the natural world.

- It develops the dispositions to use the skills, abilities, and habits of mind associated with science.

Certainly, the standards on Inquiry have direct implications for the Teaching Standards. Using inquiry to describe teaching methods shifts the focus from students' ability and understanding to the teacher's strategies and methods.

Science Subject Matter

The *Standards* relied on the conventional categories of physical, life, and earth and space science to organize major scientific concepts, principles, and theories. Although some have criticized this use of traditional divisions, the content working group found that it is not easy to create new categories that are widely accepted by both the scientific and the educational communities. Recommendations to adopt more integrated and interdisciplinary ways usually reflected a curricular orientation. In the end the committee decided that the conventional categories served as a clear reminder of the origins of fundamental conceptual schemes and the structure of the disciplines, that they had wide support, that it was not up to the national standards to reorganize content, and that they accommodated variations in curricular orientation. In most cases, conceptual organizers are varied to reflect, as closely as possible, students' developmental levels. For Physical Science, they include

Grades K–4

- properties of objects and materials
- position and motion of objects
- light, heat, electricity, and magnetism

Grades 5–8

- properties and changes of properties in matter
- motions and forces
- transformations of energy

Grades 9–12

- structure of atoms
- structure and properties of matter
- chemical reactions
- forces and motion
- conservation of energy and the increase in disorder
- interactions of energy and matter

for Life Science:

Grades K–4

- characteristics of organisms
- life cycles of organisms
- organisms and environments

Grades 5–8

- structure and function in living systems
- reproduction and heredity
- regulation and behavior
- populations and ecosystems
- diversity and adaptations of organisms

Grades 9–12

- cells

- biological evolution

- interdependence of organisms

- molecular basis of heredity

- matter, energy, and organization in living systems

- the behavior of organisms

and for Earth and Space Science:

Grades K–4

- properties of earth materials

- objects in the sky

Grades 5–8

- structure of the earth system

- earth's history

- earth in the solar system

Grades 9–12

- energy in the earth system

- geochemical cycles

- origin and evolution of the Earth system

- origin and evolution of the universe

Science and Technology

Many students do not realize that science and technology have important connections. The national *Standards* clarifies these similarities and differences. At all grade levels, the conceptual organizers for this standard are "Abilities of Technological Design" and "Understandings About Science and Technology," which were deliberately established to parallel the "Abilities of Scientific Inquiry" and "Understandings About Scientific Inquiry" in the Inquiry Standard. The other Content Standards emphasize the development of understandings.

Science in Personal and Social Perspectives

Contemporary visions of scientific literacy acknowledge the connections between science and the decisions individuals make about personal and social issues. The

Content Standard for this category should contribute to students' understanding and development of decision-making skills that will enable them to fulfill their obligations as citizens. They include

Grades K–4

- personal health

- characteristics and changes in populations

- types of resources

- changes in environments

- science and technology in challenges

Grades 5–8

- personal health

- populations, resources, and environments

- natural hazards

- risks and benefits

- science and technology in society

Grades 9–12

- personal and community health

- population growth

- natural resources

- environmental quality

- natural and human-induced hazards

- science and technology in local, national, and global challenges

The History and Nature of Science

Contemporary science builds on its history and advances knowledge through established rules of conduct, for example, using empirical evidence, applying logical argument, and allowing skeptical criticism. Standards for this category do not imply that students should understand the entire history of science, rather, they suggest and encourage the study of historical episodes such as those described in *Science for All Americans* (Rutherford and Ahlgren, 1989) and *Benchmarks for Science Literacy* (American Association for the Advancement of Science, 1993). Important figures illustrate the process of scientific inquiry, demonstrating that it usually changes very little but can sometimes have major advances, that science and society interact interdependently and that science is a fundamentally human enterprise, which has involved men and women of various cultures. The conceptual organizers for this category include

Grades K–4

- science as a human endeavor

Grades 5–8

- science as a human endeavor
- nature of science
- history of science

Grades 9–12

- science as a human endeavor
- nature of scientific knowledge
- historical perspectives

Unifying Concepts and Processes

This category describes the concepts and processes of science that connect disciplines and provide the foundation for integrated or interdisciplinary approaches to science curriculum. Its organizers include

- systems, order, and organization
- evidence, models, and explanation
- change, constancy, and measurement
- evolution and equilibrium
- form and function

Students' understandings and abilities should be developed in association with the other Content Standards over the entire K–12 continuum.

A Question

What is new and different about this vision of scientific literacy? Not very much (see, for example, Pella, 1967; Agin, 1974; Showalter, 1974; Rutherford and Ahlgren, 1989). Certainly, topics and subjects have been updated, but the basic structure is generally consistent with earlier descriptions and with the recent *Benchmarks for Science Literacy* (American Association for the Advancement of Science, 1993). This general consistency is encouraging, since it suggests that the *Standards* is in tune with the historical goals of science education. Perhaps a better question is, How does the *Standards* vision of scientific literacy compare to the content in contemporary school science programs? The answer to this question anticipates some of the implied changes in school science programs and practices.

One of the most significant changes can be found in the emphasis given to scientific inquiry. Although this goal was widely supported in the curriculum reform of the 1960s, it was not widely implemented, and where it was, it has not survived the selective pressures of the contemporary school system. Another involves the aforementioned shift in emphasis from the "processes of science" to cognitive abilities and the critical thinking associated with the development of scientific explanations.

Changes in subject matter seem relatively minor, since the *Standards* includes the basic concepts, principles, and theories associated with the science disciplines. Contributions from the scientific community and an extensive review process helped clarify the fundamental aspects of traditional conceptual schemes, such as "the structure of atoms," "the cell," and "geochemical cycles." The emphasis, however, shifts from facts, information, and vocabulary to major scientific ideas and a few related fundamental concepts.

The inclusion of technology as a domain of study constitutes a significant addition to the contemporary science curriculum. This standard was carefully designed to complement the growing interest in technology education, and Dr. Paul Black helped develop early drafts. The Science and Technology Standard presented the process of technological design as well as certain concepts associated with technology, and then applied those ideas to help students understand the scientific and technological enterprise.

The standard on Science in Personal and Social Perspectives represents one of the major differences between contemporary science education and the new vision of science literacy. The conceptual organizers associated with this standard received strong support, but debate and some conflict arose over whether they should be placed with the subject matter domains or stand alone as a separate standard. This standard is probably the most difficult for schools, since it seems the most alien to extant programs and teaching staff. Inquiry and the History and Nature of Science represent standards everyone believes in but no one does much about. Now both have been allotted position and power. Unifying Concepts and Processes had inconsistent support throughout the development of the *Standards* but survived. These standards cut across disciplines to equip students with powerful and helpful ways of thinking about their world.

A Definition of Scientific Literacy

The *Standards,* specifically the Content Standards, combined with the framework for scientific literacy described in earlier chapters, establishes a contemporary definition of scientific literacy that is clear, complete, and usable. The general goals for science education—acquiring knowledge, developing the skills and abilities associated with inquiry and design, and understanding and applying ideas and skills in personal and social contexts—have a long history (Bybee and DeBoer, 1993; DeBoer and Bybee, 1995). Science content from the national standards clarifies the dimensions—functional, conceptual and procedural, and multidimensional— and the depth of scientific literacy expected of all students who finish high school.

National Standards and the Goals of Science Education

The framework for scientific literacy in Figure 6.1 is adapted from Mortimer Adler's (1982) *The Paideia Proposal.* Each of the three columns includes distinctive goals and content orientation. Because the goals might be translated to curriculum and instruction, all three columns are essential to the development of scientific literacy. They also parallel the earlier question, What should the technologically literate person know, value, and do as a citizen? Scientific literacy includes acquiring organized knowledge, developing cognitive abilities and manipulative skills, and enlarging understanding of ideas and values through contextual issues such as history and personal and social perspectives.

The first column focuses on acquiring knowledge in six domains (all included in the *Standards*): physical science, life science, earth and space sciences, unifying concepts, scientific inquiry, and technological design. These domains, however, are not necessarily curriculum or courses of study. In fact, the science curriculum should include content from all three columns and may emphasize science-related social challenges, scientific inquiry, or unifying concepts and processes in an integrated approach.

Science education programs have traditionally done very little to help students acquire knowledge *about* science and technology, yet this knowledge is basic to many discussions, and the lack of it the cause of much confusion about the role,

FIGURE 6.1 *A Framework for the Content Standards and the Goals of Science Education*

	Goals of Science Education		
	Acquisition of organized knowledge	Development of cognitive abilities and manipulative skills	Enlarged understanding of ideas and values
	In the areas of	*In the processes of*	*In the areas of*
Content domains from the *National Science Education Standards*	▪ Physical science ▪ Life science ▪ Earth and space science ▪ Unifying concepts ▪ Scientific inquiry ▪ Technological design	▪ Scientific inquiry ▪ Technological design ▪ Unifying processes	▪ Personal matters ▪ Science-related social challenges ▪ Historical perspectives ▪ Nature of science and technology

limits, and possibilities of science and technology in contemporary society. It is vitally important for students to acquire scientific knowledge as a part of their developing scientific literacy. Some have rightly criticized the constructivist approach to science teaching as meaning that students' explanations should be accepted as scientific (Holton, 1993). What such a view fails to recognize is that a body of knowledge called science exists and that students' explanations cannot simply be labeled scientific. To view scientific literacy in this way is detrimental.

The second column emphasizes the abilities and skills associated with scientific inquiry and technological design. As I have pointed out, the national standards for Science as Inquiry and for Science and Technology elaborate the specific abilities related to inquiry and design. If we use scientific inquiry as an example, the related abilities and skills include applying science processes (observing, inferring, experimenting, classifying, controlling variables), constructing scientific explanations (interpreting data, using critical thinking and logic to link evidence to explanation, formulating models, defining operationally), recognizing alternative explanations (maintaining an open mind, accepting the tentative nature of explanations, being skeptical), and communicating (reading, writing, speaking, listening). If the first column is "know about," the second is "know how." In column two, the learning outcomes indicate that students know how to do scientific investigations. What is innovative about these goals is that students engage in inquiry and design. Most evidence indicates that students do not have many such experiences in their science education.

The content in the third column provides important contexts for teaching science and for learning science. Scientific literacy locates science in the context of history, society, and individual decisions. The content domains in column three engage students in ways that produce meaningful knowledge and sharpen cognitive skills. As students encounter science in various historical, personal, and social contexts, they come to recognize its ideas and values and further enlarge them in their own learning. The *Standards* placed significant value on the development of understanding, which occurs when students have to respond to the challenges of science and technology in deciding issues in their own lives, in considering societal problems, and in evaluating historical perspectives. The content encourages scientific literacy because these topics require students to practice their cognitive skills and draw on their scientific knowledge as they analyze various positions. They also have an opportunity to correct misconceptions and enlarge their own ideas.

The National Standards and the Framework for Scientific Literacy

Achieving scientific literacy is a lifelong process. In many discussions, scientific literacy is seen as a single goal that one either attains or does not attain: "Students could not identify the chemical structure of DNA" and "they could not explain the reasons for seasons"—so "we have a nation of scientifically illiterate citizens." Much more helpful than these either/or absolutes is the view that all students (and adults, for that matter) occupy positions somewhere on a literacy continuum, which includes vocabulary, concepts, capacities, and a rich and integrated multidimensional understanding of science and technology. It also seems clear that scientific words and concepts are associated with different levels or depths of understanding. A student might correctly spell and use the word *cell* in a simple sentence but fail to recognize other important aspects about cells, for example, that they convey information and that they reproduce. If asked about the function of cells in cancer, the student has no idea about the relationship. Is this student scientifically literate? Using a strict definition of "literacy," that is, the ability to read and write the word *cell,* one would have to say the student is scientifically literate for this life science topic. What about the fact that the same student does not understand the structure and function of cells or the fact that cells play a role in cancer? It might be easier to use an either/or approach, but it will not be very helpful in deciding what to do to further this student's understanding of science.

The following dimensions of scientific literacy should not be interpreted as developmental stages or instructional sequences but rather, as different aspects of a constellation of knowledge, ability, skill, and understanding referred to as *scientific literacy.* The national standards attempt to supply answers for questions such as, What should I teach about "chemical reactions," "matter, energy, and organization in living systems," or "the origin and evolution of the earth system"? Whether it is a technical word such as *ozone;* a conceptual organizer such as "interactions of matter and energy"; or a fundamental axiom such as "any ordered state tends to become disordered over time," the process of designing curriculum materials and teaching encourages greater depth.

SCIENTIFIC ILLITERACY   The idea of scientific *literacy* presupposes scientific *illiteracy.* What I mean is that some individuals, because of age, stage of development, or impaired cognitive abilities, may be scientifically illiterate. They are unable to respond to reasonable questions about science, nor do they have the vocabulary, concepts, contextual background, or cognitive capacity to identify a question or topic as scientific. In contrast, there are those who proclaim that scientific illiteracy is widespread based on a small sample and specific assessment. Such proclamations may gain media attention, but they have little educational value. Clearly, however, the perceptions conveyed by language make a difference. There is a difference between focusing on the large number of individuals who are alleged to be scientifically illiterate because they cannot, for example, explain the difference between *mass* and *weight, heat* and *temperature,* or the earth's *rotation* and *revolution,* and the view that all individuals fall somewhere on a continuum of scientific literacy: their knowledge and abilities vary with subject matter and context. My educational position is one of helping individuals develop those dimensions where they lack depth and understanding. This is a much more educationally sound view of scientific literacy, and one that science teachers and science education can act on.

NOMINAL SCIENTIFIC LITERACY   A science teacher begins a lesson on *force* and soon discovers that students think force is a property of a moving object. The teacher then tries to introduce the idea that forces are characteristics of action between objects, but the students' misconceptions persist. This example is from *Making Sense of Secondary Science* (Driver et al., 1994), and exemplifies nominal scientific literacy in relation to the concept of force. The students understood the topic as scientific, but their level of understanding reveals a misconception. To summarize, at the nominal level of scientific literacy, the learner associates names and ideas in a general way with science and technology, but the association represents a misconception, naive theory, or inaccurate concept. The relationship between these terms and concepts and acceptable scientific definitions is small and significant. The learner has, at best, only a token understanding.

The extensive body of research on misconceptions demonstrates that nominal scientific literacy is crucial to achieving higher levels of scientific literacy. In most instances, this is the point where teaching and learning begin. But the value of this view, and the associated research paradigm, extends beyond discussions of students' misconceptions. Our perceptions of teaching and learning have shifted, from the "empty vessel" model to one that involves replacing misconceptions with more scientifically accurate concepts. This identification of a dimension of nominal scientific literacy is consistent with wider discussions of constructivism and with current understanding of learners and the contemporary learning model (American Psychological Association, 1993).

The *Standards* acknowledges the presence of nominal scientific literacy in "Developing Student Understanding," a section that immediately follows each content standard and gives examples of the types of knowledge, the abilities, and the level of understanding that educators might encounter. Given the nature of the *Standards,* a research review of all misconceptions associated with each standard would have been impossible.

FUNCTIONAL SCIENTIFIC LITERACY Learners demonstrating functional scientific literacy respond adequately and appropriately to the vocabulary associated with science and technology and meet minimum standards of literacy, but they demonstrate little knowledge of scientific concepts, principles, laws, or theories and the fundamental procedures and processes of scientific inquiry. Learners may be familiar with scientific terms through science classes, visits to museums, television, or books. Textbooks and programs that emphasize rote memorization exclusively encourage functional levels of scientific literacy but leave learners with little or no understanding of the disciplines, no experience of the excitement of inquiry, and probably little interest in science. Tables 6.1, 6.2, and 6.3 summarize some of the technical terms from the *Standards*. I have selected the words from the conceptual organizers and fundamental understandings for all content standards I thought essential to communicating scientific concepts or processes. In general, I avoided common words, such as *water* and *air,* but in some cases, I incorporated otherwise common words, such as *alternative explanations,* because they had a specific meaning in the standard. In other cases, I found it necessary to include particular phrases, such as a "way of knowing," because they also had specific meaning for science and science education. Finally, even though the standard on Unifying Concepts and Processes was designed to span the K–12 continuum, I found it necessary, and probably helpful to science teachers and curriculum designers, to identify some vocabulary for this standard at K–4, 5–8, and 9–12.

The development of the Content Standards did not begin with discussions of vocabulary. In fact, the vocabulary listed in Tables 6.1, 6.2, and 6.3 emerged during discussions of concepts as the research group struggled to express ideas clearly and nontechnically. Although the list of vocabulary may seem extensive, it is a truncated version of that in most current elementary, middle, and high school science textbooks. Yet vocabulary represents only one dimension of scientific literacy; the technical words listed here constitute the core knowledge in the *Standards* and reinforce connections to other dimensions of scientific literacy.

CONCEPTUAL AND PROCEDURAL SCIENTIFIC LITERACY Students develop some understanding of the major conceptual schemes of science in general and of those related to specific disciplines. In the *Standards,* this dimension is expressed primarily in the traditional domains of the physical, life, and Earth and space science and through the processes of inquiry and design. Learners would, for example, begin to understand the central ideas of matter, energy, and motion in the physical sciences, evolution in the biological sciences, and geochemical cycles in the Earth sciences and identify how new explanations and inventions emerge.

Procedural literacy includes the processes of scientific inquiry and technological design: learners have the requisite ability, and they understand that technological design includes identifying appropriate problems, designing a solution, implementing the solution, evaluating the design or product, and communicating their discoveries and conclusions. Learners understand the structure of scientific disciplines and the procedures for developing new knowledge and techniques.

In Figures 6.2 through 6.6, I have reorganized and summarized the conceptual organizers and fundamental concepts from grades 9–12 in the *Standards* into a framework of scientific literacy because they will eventually answer questions about the degree to which students are achieving scientific literacy. The statements in the figures convey the essential concepts and processes outlined in the national standards. However, in developing the figures, it was important to "unpack" some of the fundamental concepts, and I provide an educational rationale for doing so. In some cases, the conceptual "load" of a statement was quite high, and from the point of view of curriculum, instruction, and assessment, the division is intended to be helpful to those who use this framework.

MULTIDIMENSIONAL SCIENTIFIC LITERACY These aspects of scientific literacy incorporate an understanding of the sciences that extends beyond the concepts of the scientific disciplines and the procedures of scientific investigation to include the philosophical, historical, and social dimensions of science and technology. Learners begin to make connections within scientific disciplines, between

FIGURE 6.2 *Conceptual Scientific Literacy: Inquiry*

Conceptual organizer

Understandings about Scientific Inquiry

Fundamental concepts

- Scientists usually base their inquiries on existence questions or causal-functional questions.
- Scientists conduct investigations for a variety of reasons: exploring new areas, discovering new aspects of the natural world, confirming prior investigations, predicting based on current theory, and comparing models and theories.
- Scientists rely on technology to enhance data gathering and data manipulation.
- Scientific explanations must adhere to criteria, such as logical structure and rules of evidence, be open to criticism and modification, connect to historical and current knowledge base, and report methods and procedures used to obtain evidence.
- Scientists communicate and defend the results of their inquiries: new knowledge and methods.

---

science and technology, and between science and technology and the larger issues of societies and cultures. Although a number of individuals have presented frameworks for scientific literacy that incorporate some of these dimensions, two examples dominate the contemporary science education, the *Standards* and the *Benchmarks*. The essential view of scientific literacy in both documents is similar and consistent with the framework I am proposing. The multidimensional perspective of scientific literacy has two components, one that integrates various aspects of scientific concepts and procedures and one that enlarges understanding through the study of science within the contexts of other disciplines, society, and history. *Integral* literacy means understanding the essential conceptual structures of science as well as the features that make that understanding more complete. In the *contextual* dimension, learners see the relationship of disciplines to the whole of science and technology and to various personal issues and societal challenges. The essential feature of multidimensional scientific literacy is that learners have an enlarged understanding of the concepts, processes, and values of science. Connections, contexts, and completeness express the spirit of this perspective.

In reorganizing the national standards into this framework, I decided that the standard on Science in Personal and Social Perspectives described the conceptual dimension, while standards on Science and Technology and the History and Nature of Science better represented the multidimensional category. Figures 6.7 and 6.8 present the conceptual organizers and fundamental concepts from grades 9–12 in the national standards for multidimensional scientific literacy.

TABLE 6.1 National Science Education Standards *Functional Scientific Literacy (Grades K–4)*

Content Standard	*Technical Words for Scientific Literacy*				
Science as inquiry	Asking questions Classifying Communication	Describing Evidence Fair test	Information Instrument Investigation	Observation Scientific	
Physical science	Absorbed Attract Circuits Color Electricity Gas	Heat Light Liquid Magnetism Materials Metal	Motion Objects Paper Pitch Position Properties	Reaction Reflected Refracted Repel Shape Size	Solid Sound Temperature Vibration Weight Wood
Life science	Animals Basic needs Behavior Birth	Death Depend Environment Functions	Growth Inherit Life cycle Microorganisms	Organisms Plants Senses	Structures Survive
Earth and space science	Atmosphere Clouds Earth Earthquake Erosion	Fossils Landslide Moon Precipitation Resources	Revolution Rocks Rotation Season Sky	Soil Stars Sun Volcano Weather	Weathering Wind
Science and technology	Constraints Cost Designed (objects)	Invention Materials Natural (objects)	Problem Safety Science Solution	Space Team Technology Time	Tools
Science in personal and social perspectives	Communication Disease Health	Nutrition Population Resources	Safety Sanitation Security	Substances Transportation	
History and nature of science	Contribution (to science)	History			
Unifying concepts and processes	Balance Change Constancy	Explanation Form Function	Interactions Measure Model	Order Organization Predict	System

Transforming the Vision to Reality

States, communities, and schools should not place the entire burden for these improvements in achieving scientific literacy solely on science teachers in elementary, middle, and high schools, although clearly, science teachers will have to change so that students can meet the goals of this vision of scientific literacy. This is not only an era of standards-based reform, it also is an era of systemic reform. The

TABLE 6.2 National Science Education Standards *Functional Scientific Literacy (Grades 5–8)*

Content Standard	*Technical Words for Scientific Literacy*			
Science as inquiry	Alternative (explanations) Analyze Cause-Effect	Control Critique Data Explain	Interpret Logic Propose (explanations) Reasoning	Skepticism Variables
Physical science	Acids Boiling point Chemical reaction Compound Conservation Constant	Density Direction Electrical Elements Forces Infrared	Magnitude Mass Mechanical Nuclear Radiation Solubility	Speed Transformation Transmission Ultraviolet Wave length
Life science	Abiotic Adaptation Asexual Bacteria Biotic Cell Chromosomes Circulation system Consumers Control system Coordination system Decomposers Digestion system	Diversity Ecosystem Eggs Evolution Excretion system Extinction Food web Fungi Genes Heredity Infection Levels of organization Limiting factors Movement system	Multicellular Niches Organ Organ systems Photosynthesis Physiology Population Predator Prey Producers Protection system Regulation Reproduction system	Reproduction Respiration system Response Selection Sexual Species Sperm Tissue Trait Variation
Earth and space science	Asteroid Atmosphere Catastrophe Climate Comet Condense Constructive forces Continents Core Crust	Crustal plates Deposition Destructive forces Earth system Eclipse Erosion Evaporates Global Gravity Local	Mantle Mountain building Nitrogen Orbit Oxygen Phases of moon Planets Recrystallize Rock cycle Sediments	Solar system Solvent Texture Tides Trace gases Water cycle Weathering

educational system includes all those in the science education community, and achieving the vision and goals of scientific literacy requires leadership from the entire community.

This vision of scientific literacy answers the question: What should a scientifically literate person know and do—as a citizen? In answering this question, the *Standards* had to "create a vision of the scientifically literate person" and

Content Standard	Technical Words for Scientific Literacy			
Science and technology	Aesthetics Back-up system Benefit Constraint	Criteria Efficiency Product Risk	Side-effects Techniques Temporary Time	Trade-offs Unintended consequences
Science in personal and social perspectives	Abused substances Addiction Alcohol Beneficial Biological hazards Cardiovascular endurance Chemical hazards Consent Detrimental	Differential effects Economic growth Environmental degradation Ethical code Exercise Harmful Long-term Mental health Natural hazards	Overconsumption Overpopulation Physical fitness Prior knowledge Probability Rate of change Resource acquisition Resource depletion Risk analysis Scale of change	Sex drive Short-term Social hazards Social needs Tobacco Transmitting disease Urban growth Waste disposal World view (of science)
History and nature of science	Confirmation Creativity Engineering Evaluation Experiment	Habits of mind Honesty Insight Interpretation Mathematical model	Openness to ideas Phenomenon Reasoning Skepticism Tentative	Theoretical model Theory Tolerance of ambiguity
Unifying concepts and processes	Cause-effect Closed system Cycle Evidence for interaction Feedback Flow of resources (system) Hierarchies	Law Levels of organization Open system Paradigm Pattern of change Probability Qualitative description	Quantitative description Rate of change Regularities in systems Scale of change Subsystem System boundaries System components	System properties Theory Travel

describe the knowledge, skills, and understandings aligned with this vision. Making that vision a reality asks science educators and science teachers to understand the various domains and dimensions in the framework for scientific literacy, and then to begin to address issues related to curriculum, instruction, and assessment.

Opportunities to Learn

The aforementioned framework defines scientific literacy, but it does not describe a course of study. From a science teacher's point of view, the standards are not lessons, units, classes, or a science curriculum. In developing the national standards,

TABLE 6.3 National Science Education Standards *Functional Scientific Literacy (Grades 9–12)*

Content Standard	*Technical Words for Scientific Literacy*		
Science as inquiry	Analysis of data Causal-functional questions Display of data Logical argument Open to question	Plausible explanation Public communication Rules of evidence Science concepts	Scientific argument Statistical analysis Testable hypothesis
Physical science	Absorb light Accelerate Acid base reaction Atomic number Atomic interactions Bond Carbon atoms Carbon-based molecules Catalysts Chemical compound Chemical properties Conduction Conservation of insulating materials Convection Decay Decelerate Discrete amounts (gain/lose energy) Electric force Electromagnetic force Electromagnetic waves Emit light Energy conservation Energy transfer	Enzymes F=MA Femto seconds Field energy Fission Fusion Gamma rays Gravitational force Greenhouse gases Heat energy Increase in disorder Radioactive Inversely proportional Ions Isotopes Kinetic energy Microwaves Net force Neutrons Nuclear force Nuclear reaction Nucleus Oxidation/reduction reactions Ozone	Periodic table Physical properties Potential energy Proportional Protons Radiation Radicals Radio waves Radioactive isotopes Reacting species Reaction rates Semiconducting material Share electrons Smog Square of the distance Strength of force Superconductors Synthetic polymers Transfer Visible light Waves X-rays
Life science	Anthropology ATP Biological classification Biosphere Carbon dioxide Carnivores Cell differentiation Compete Cooperate	Embryo Encoded Energy production Excitatory molecules Finite Generation Genetic information Genetic variability Habitat	Interrelationship Membrane Mutations Natural selection Nerve cells Nervous system Organic molecule Phosphate bond Pollution

Content Standard	Technical Words for Scientific Literacy		
Life science (continued)	Covalent chemical bond Descent from common ancestors DNA Recycle Regulation Sociology Synthesis of molecules Templating mechanisms X-chromosome Y-chromosome	Harvesting Herbivores Infinite Inhibitory molecules Interdependence	Progeny Protein synthesis Protein catalysts–enzymes Psychology Recombination of genes
Earth and space science	Big Bang theory Correlate External sources of energy Galaxies	Geochemical cycles Geologic time Global climate Human scale Hydrogen	Internal sources of energy Nebular cloud Reservoirs Rock sequences Universe
Science and technology	Artifacts Interface Patents Public knowledge	Private knowledge Simulations Science disciplines	
Science in personal and social perspectives	Birth rate Cancer Carrying capacity Consumption Cultural norms Death rate Ethics	Exponential growth Family system Fertility rate Infant mortality rate Land use Level of affluence Linear growth	Nutritional balance Production Religious beliefs Renewable resource Resistance (to disease) Sexuality Virulence
History and nature of science	Certainty Consistency Empirical standards Ethical tradition (of science) Logical argument	Openness to criticism Rules of evidence Skeptical review Way(s) of knowing	
Unifying concepts and processes	Equilibrium Evolution Explanatory model Homeostasis Input	Negative feedback Output Positive feedback Relative certainty (or uncertainty) Steady state Unit of analysis	

FIGURE 6.3 *Conceptual Scientific Literacy: Physical Science*

Conceptual organizers

	Structure of atoms	Structure and properties of matter	Chemical reactions
Fundamental concepts	▪ Matter consists of minute particles called atoms, and atoms are composed of even smaller components. ▪ The atom's nucleus is composed of protons and neutrons, which are more massive than electrons. ▪ Nuclear forces that hold the nucleus together, at nuclear distances, are usually stronger than the electric forces that would make it fly apart. ▪ Radioactive isotopes are unstable and undergo spontaneous nuclear reactions, emitting particles and/or wave-like radiation.	▪ Atoms interact with one another by transferring or sharing electrons that are farthest from the nucleus. These outer electrons govern the element's properties. ▪ An element is composed of a single type of atom. ▪ Bonds between atoms are created when electrons are transferred or shared. ▪ Properties of compounds reflect the nature of the interactions among its molecules, and these are determined by the structure of the molecule. ▪ Solids, liquids, and gases differ in the magnitude and direction of focus between molecules or atoms. ▪ Carbon atoms can bond to one another in chains, rings, and branching networks to form a variety of structures.	▪ Chemical reactions occur all around us. ▪ Complex chemical reactions involving carbon-based molecules constantly take place in cells. ▪ Chemical reactions may release or consume energy. ▪ A large number of reactions involve the transfer of either electrons (oxidation/reduction reactions) or hydrogen ions (acid/base reactions) between reacting ions, molecules, or atoms. ▪ In some reactions, chemical bonds are broken by heat or light to form very reactive radicals with electrons available to form new bonds. ▪ Chemical reactions can occur in time periods ranging from femto seconds to geologic time. ▪ Reaction rates depend on how often reacting atoms and molecules encounter one another, the temperature, and the properties of the reacting species. ▪ Catalysts accelerate chemical reactions.

FIGURE 6.3 *Conceptual Scientific Literacy: Physical Science*

Conceptual organizers

Forces and motions

- Objects change their motion only when a net force is applied.
- Gravity is a universal force that exists on any other mass.
- Electrical force is a universal force that exists between any two charged objects.
- Between any two charged particles, electrical force is vastly greater than gravitational force.
- Electricity and magnetism are two aspects of a single electromagnetic force.

Conservation of energy and the increase of disorder

- The total energy of the universe is constant.
- All energy can be considered as kinetic—the energy of motion; potential—the energy of relative position; or field—the energy contained by a field such as electromagnetic waves.
- Heat energy consists of random motion and the vibration of atoms, molecules, and ions.
- Energy tends to move spontaneously from hotter to colder objects by conduction, convection, or radiation.
- Any ordered states tend to become disordered over time.

Interactions of energy and matter

- Waves carry energy and can interact with matter.
- Electromagnetic waves results when a charged object is accelerated or decelerated.
- Each kind of atom or molecule can gain or lose energy only in particular discrete amounts and thus can absorb and emit light only at wavelengths corresponding to these amounts.

FIGURE 6.4 *Conceptual Scientific Literacy: Life Science*

Conceptual organizers

	The cell	The molecular basis of heredity	Biological evolution
Fundamental concepts	▪ Cells have particular structures that underlie their functions. ▪ Most cell function involves chemical reactions. ▪ Cells store and use information to guide their functions. ▪ Cell functions are regulated. ▪ Plant cells contain chloroplasts, the site of photosynthesis. ▪ Cells can differentiate, and complex organisms can develop from the differentiated progeny of cell division. ▪ Cell differentiation is controlled through the expression of different genes.	▪ DNA carries the instructions for specifying the characteristics of the organism. ▪ DNA is a large polymer. ▪ The chemical and structural properties of DNA explain how the genetic information that underlies heredity is both encoded in genes and replicated. ▪ Most of the cells in a human contain two copies of each of 22 chromosomes, an additional two X chromosomes in females, or additional X and Y chromosome in males. ▪ Transmission of genetic information to offspring occurs through egg and sperm cells that contain only one representative from each chromosome pair. ▪ An egg and sperm unite to form a new human. ▪ The human is formed from cells that contain two copies of each chromosome and therefore two copies of each gene. ▪ Mutations in DNA occur spontaneously at low rates. ▪ Most mutations have no effect on organisms, however, some can change cells and organisms.	▪ Species change over time. ▪ Evolution is the consequence of interactions of (1) the potential for a species to increase its numbers, (2) the genetic variability of offspring due to mutation and recombination of genes, (3) the finite supply of resources required for life, and (4) the ensuing selection by the environment of those offspring better able to survive and leave offspring. ▪ The diversity of life is a result of more than 3.5 million years of natural selection and evolution. ▪ Natural selection and its evolutionary consequences provide scientific explanation for the fossil record as well as the similarities of molecular structure among diverse species of living organisms. ▪ The millions of species on earth are related by descent from common ancestors. ▪ Biological classifications indicate how organisms are related and their evolutionary relationships.

FIGURE 6.4 *Conceptual Scientific Literacy: Life Science*

Conceptual organizers

The interdepence of life

- Atoms and molecules on earth cycle among living and nonliving components of the biosphere.
- Energy flows through ecosystems in one direction, from photosynthetic organisms to herbivores to carnivores and decomposers.
- Organisms both cooperate and compete in ecosystems.
- Living organisms have the capacity to produce populations of infinite size, but environments and resources are finite.
- Humans live within the world's ecosystems.

Matter, energy, and organization in living systems

- All matter tends toward more disorganized states.
- Living systems require a constant input of energy.
- The energy for life ultimately derives from the sun.
- Plants capture light energy and use it to form strong (covalent) chemical bonds between the atoms of carbon-containing organic molecules.
- The chemical bonds of food molecules contain energy.
- The complexity and organization of organisms accommodates the need for obtaining, transforming, transporting, releasing, and eliminating the matter and energy used to sustain the organism.
- The distribution and abundance of organisms and populations in ecosystems are limited by the availability of matter and energy and the ability of the ecosystem to recycle organic material.
- As matter and energy flow through different levels of organization of living systems and between living systems and the physical environment, chemical elements are transformed and recombined in different ways. Each transformation results in storage or dissipation of energy into the environment as heat. Matter and energy are conserved in each transformation.

The behavior of organisms

- Multicellular animals have nervous systems to generate behavior.
- Organisms have behavioral responses to internal changes and to external stimuli.
- Behaviors have evolved through natural selection.
- Behavioral biology has implications for humans, providing links to psychology, sociology, and anthropology.

FIGURE 6.5 *Conceptual Scientific Literacy: Earth and Space Science*

	Conceptual organizers			
Fundamental concepts	*Energy in the earth System*	*Geochemical cycles in the Earth system*	*The origin and evolution of the Earth system*	*The origin and evolution of the Universe*
	▪ Earth systems have both internal and external sources of energy, both of which create heat. ▪ The major sources of internal heat energy are decay of radioactive isotopes and gravitational energy from the earth's original formation. ▪ The outward transfer of earth's internal heat drives convection circulation in the mantle resulting in movement of crustal plates. ▪ Heating of the earth's surface and atmosphere by the sun drives convection within the atmosphere and oceans, producing winds and ocean currents. ▪ Global climate is determined by the heat transfer from the sun at and near the earth's surface.	▪ The earth is a system containing essentially a fixed amount of each stable chemical atom of element. ▪ Each element exists in and moves among different chemical reservoirs, such as solid earth, oceans, atmosphere, and biosphere. ▪ Movement of matter between reservoirs is driven by the earth's internal and external sources of energy. ▪ Movement of matter between reservoirs often results in physical and chemical changes in the properties of matter.	▪ The sun, the earth, and the rest of the solar system formed from a nebular cloud of dust and gas 4.6 billion years ago. ▪ The early earth was very different from the planet we live on today. ▪ Geologic time can be estimated using several methods including observing rock sequences, correlating fossils, and decay rates of radioactive isotopes in rock formations. ▪ Evolution of the earth system results from interactions among lithosphere, hydrosphere, atmosphere, and biosphere. ▪ Some changes in the earth system occur on a human scale and others occur on a geologic scale. ▪ Evolution of life resulted in dramatic changes in the composition of the earth's atmosphere, which did not originally contain oxygen.	▪ The Big Bang theory places the origin between 10 and 20 billion years ago, when the universe began in a hot dense state and has been expanding every since. ▪ Trillions of stars were formed early in the history of the universe when matter, primarily the light atoms hydrogen and helium, clumped together due to gravitational attraction. ▪ Billions of galaxies, each of which is a gravitationally bound cluster of billions of stars, now form most of the visible mass in the universe. ▪ Stars produce energy from nuclear reactions, primarily the fusion of hydrogen to form helium. ▪ Nuclear reactions and other processes also produce all other elements.

FIGURE 6.6 *Procedural Scientific Literacy: Scientific Inquiry and Technological Design*

	Procedural organizers	
	Abilities of scientific inquiry	*Abilities of technological design*
Fundamental abilities	▪ Identify questions and concepts that guide scientific investigations. ▪ Design a scientific investigation. ▪ Use technology to improve investigations and communications. ▪ Formulate and revise scientific explanations and models using logic and evidence. ▪ Recognize and analyze alternative explanations and models. ▪ Communicate and defend a scientific argument. ▪ Cell differentiation is controlled through the expression of different genes.	▪ Identify a problem or design opportunity. ▪ Propose designs and choose between alternative solutions. ▪ Implement a proposed solution. ▪ Evaluate the solution and its consequences. ▪ Communicate the problem, process, and solution.

we presented content to represent inquiry and the domains of the physical, life, and earth sciences and accommodated the historically important goals of science education (Bybee and DeBoer, 1993; DeBoer and Bybee, 1995).

Science teachers and curriculum developers can organize science content according to a variety of topics, perspectives, and emphases. One can imagine a curriculum emphasis that includes Scope, Sequence, and Coordination (National Science Teachers Association, 1992), STS (Yager, 1993), or the conceptual schemes of the science disciplines, such as biology (Biological Sciences Curriculum Study, 1993). However, these curriculum orientations may not provide adequate opportunities for students in all the standards, so school science programs may have to include a variety of curriculum emphases and instructional approaches.

This vision of scientific literacy implies a greater emphasis on inquiry, technology, science in personal and social perspectives, and the history and nature of science. But the vision represents only one part of the task. A second task involves providing adequate opportunities for students to learn the content, which will require new strategies and methods of teaching as well as a reconsideration of what is important. One example of the shift is inquiry. Students come to understand the nature of science through their active involvement in inquiry-oriented investigations: they need to engage in investigations, and identify the questions and concepts that guide those investigations, design investigations, use critical thinking and logic to make connections between evidence, science knowledge, and their explanations of phenomena. Curriculum developers will have to design and develop programs that science teachers will undoubtedly have to implement exemplary teaching practices and employ new and different technologies.

FIGURE 6.7 *Multidimensional Scientific Literacy: Integral*

Understandings about Science and technology	*Science as a human endeavor*	*Nature of scientific knowledge*	*Historical perspectives*
■ Scientists in different disciplines ask different questions, use different methods of investigation, and accept different types of evidence to support their explanations. ■ Science often advances with the introduction of new technologies, and solving technological problems often results in new scientific knowledge. ■ Creativity, imagination, and a good knowledge base are all required in the work of science and engineering. ■ Scientific inquiries are pursued by the desire to understand the natural world, and technological designs are pursued by the need to meet human needs and solve human problems. ■ Science answers questions that may or may not directly influence humans and technology has a more direct effect on society because it solves human problems, helps humans adapt, and fulfills human aspirations. ■ Scientific knowledge is made public, and technological knowledge is often not made public because of patents and the financial potential of an idea, invention, or product.	■ Individuals and teams have contributed and will continue contributing to the scientific enterprise. ■ Individuals pursue science and engineering as careers. ■ Scientists abide by ethics and rules, such as commitment to peer review, truthful reporting, and making knowledge public.	■ Science distinguishes itself from other ways of knowing and bodies of knowledge through the use of empirical standards, logical arguments, and skeptical review. ■ Scientific explanations must meet criteria, such as consistency with experimental and observational evidence, accurate predictions, connections with extant knowledge, and logical structures. ■ Scientific knowledge is, in principle, subject to change as new evidence becomes available.	■ Historically, diverse cultures have contributed scientific knowledge and technologic inventions. ■ Usually, advances in science and technology occur as small modifications in the current knowledge base and techniques and products to meet human needs and aspirations. ■ Occasionally, there are advances in science and technology that have long-lasting and widespread effects.

FIGURE 6.8 *Multidimensional Scientific Literacy: Contextual*

Personal and community health

- Hazards and the potential for accidents exist.
- Some diseases are caused by microorganisms and some are caused by specific body dysfunctions.
- Personal choice concerning fitness involves multiple factors.
- An individual's mood and behavior may be modified by substances that have beneficial or detrimental effects.
- Drugs can result in physical dependence and increase the risk of injury, accidents, and death.
- Selection of foods and eating patterns determine nutritional balance that has a direct effect on growth and development and personal well-being.
- Family systems serve basic health needs.
- Sexuality is basic to the physical, mental, and social development of humans.

Population growth

- Populations grow or decline through the combined effects of birth and deaths, and in countries, through emigration and immigration.
- Factors, such as levels of affluence and education, importance of children in labor force, education and employment of women, infant mortality rates, cost of raising children, availability and reliability of birth control methods, religious beliefs, and cultural norms, influence personal decisions about family size and subsequently birth rates and fertility rates.
- Populations can reach the limits to growth or carrying capacity of the environment and resources.

Natural resources

- Human populations use resources and the environment in order to maintain and improve their existence.
- The earth does not have infinite resources, increasing production and consumption places severe stress on resources and the processes that renew some resources.
- Humans use many natural systems as resources.

Determining the Degree to Which We Achieve the Vision

The third, and most essential, task is assessment, establishing what counts as attaining the understanding and abilities described in the *Standards*. Here again we have a defining quality of scientific literacy. For example, science teachers have to assign end-of-term grades, and to do so, they have to decide what will count as the measure of successful achievement of scientific literacy. They will need to use a variety of ways to obtain information about student achievement including techniques such as portfolios and performance assessments.

FIGURE 6.8 *Multidimensional Scientific Literacy: Contextual (continued)*

Environmental quality

- Natural ecosystems provide an array of basic processes that affect humans.
- Materials from human societies disturb both physical and chemical cycles of the earth.
- Many factors influence environmental quality.

Natural and human-induced hazards

- Normal adjustments of earth may be hazards for humans.
- Human decisions and activities can enhance potential for hazards.
- Some hazards, such as earthquakes, volcanoes, and severe weather, are rapid and spectacular.
- Natural hazards present the need for humans to assess potential danger and risk.

Science and technology in local, national, and global challenges

- Science and technology are essential social enterprises, but alone they can only indicate what can happen, not what should happen; the latter involves human decisions.
- Understanding the basic concepts and principles of science and technology should precede active debate about the economics, politics, and ethics of various decisions.
- Progress in science and technology can relate to social priorities, issues, and challenges.
- Decisions about science and technology and social challenges involve assessment of alternatives, risks, costs, consideration of who benefits, who pays, and who assumes the risks.
- Individuals and society should ask: "What can happen?" "What are the odds?" and "How do scientists and engineers know what will happen?" concerning science- and technology-related social challenges.

Conclusion

The *Standards* represents a broad effort to improve scientific literacy. It is easily the most significant document in science education of this decade and prominent among those on the short list of documents in the history of science education. The fact that the standards for science education are part of a standards-based reform that incorporates other school subjects guarantees support from the larger educational community for the improvement of science content, teaching, and assessment. The challenge of understanding and implementing the vision of scientific literacy is important, and achievable. It requires change—for science teachers, as well as school administrators, scientists and engineers, legislators, and parents—all of whom should provide continuing support for the improvement of science education.

7 *Enriching the Science Curriculum*

Publication of *Benchmarks for Science Literacy* (American Association for the Advancement of Science, 1993) and *National Science Education Standards* (National Research Council, 1996a) signaled a new era of curriculum reform. In terms of the themes of this book, these documents moved the discussion from policies to programs and practices. Since the early 1980s, numerous reports have suggested the need for reform, and many legislatures and school districts have changed policies regarding such issues as graduation requirements and length of school days and academic years. But until the early 1990s, science educators had not adequately addressed the issue of curriculum reform and school science programs. With publication of national standards and benchmarks, and the widespread revision of state and local frameworks, attention has turned to the improvement of the science curriculum and the associated issues of improved classroom practices.

Perspectives on the Development of the Science Curriculum

This section presents my perspective as a science curriculum developer, notably at the BSCS. My ideas did not originate in curriculum theory; rather they went from practice to theory, not the other way around. My perspective originated in the practical arena of developing science curricula for elementary schools, middle schools, high schools, and colleges. In many important respects, the perspective expressed here complements other discussions about curriculum development at BSCS (Mayer, 1976; McInerney, 1986/1987) and other places (Tyler, 1949; Schwab, 1962; Grobman, 1969; Roberts, 1980).

Dimensions of the Science Curriculum

Individuals often equate categories of science content with the scope, sequence, and emphasis of a science curriculum. In the *National Science Education Standards*

(National Research Council, 1996a), we used categories, such as physical, life, earth and space science, inquiry, technology, personal and social perspectives, and history and nature of science, to organize concepts and fundamental understandings. Translating science content standards to a science curriculum requires understanding the dimensions of the science curriculum.

The term *curriculum* covers a variety of things, such as syllabi, textbooks, materials, and school science programs. Mostly, *curriculum* signals the individual's perspective. For example, science teachers use the term to mean what they teach in ninth grade Earth science, and administrators may associate the term with textbooks and materials for elementary, middle, and high school science. Examining definitions of *curriculum* provides some, but not a lot of help, for instance, "Curriculum is all of the experiences children have under the guidance of teachers" (Caswell and Campbell, 1935), or "Curriculum encompasses all learning opportunities provided by the school" (Saylor and Alexander, 1974). Taking the time to analyze the difference between these definitions is an important academic exercise but not very helpful when one is pressed for time and must improve a particular science curriculum, usually according to a set of criteria. For instance, materials for high school physical science must be developed with NSF support over three years, or a local school district has one year to develop a philosophy statement, a district syllabus, and a supply of materials for grades K–12. This perspective is Joseph Schwab's (1970) "practical" view of curriculum making, but he also recognizes the "impractical" position of each science teacher making his or her own curriculum.

There are several ways to think about science curriculum. First, use of a "technology and design" metaphor rather than "science and inquiry" promotes an accurate perspective on the process of curriculum development that most individuals experience. Second, curriculum policy can be used to differentiate between the various standards, benchmarks, syllabi, and other criteria at national, state, and local levels, and the curriculum that is translated into school programs and practices. Third, curriculum includes more than physical materials, such as textbooks, software, and kits. Finally, there are different aspects of curriculum, such as the recommended, written, taught, assessed, and learned.

What do I mean by science curriculum? Although a science curriculum includes science content, materials, teaching methods, and learners in complex ways, one can identify an idea that unifies the complex interactions. The science curriculum represents a series of constructed relationships among conceptual schemes, procedural strategies, and contextual factors. In short, there are science concepts, processes, and topics. These constructed relationships represent the science concept and the various emphases that textbook writers, curriculum developers, teachers, and assessment specialists give to the concepts, procedures, and content. This definition varies from the common view of curriculum as an operational plan that guides learning, courses of study, and the experiences of learners. Such definitions are somewhat limited in that they usually are reduced to syllabi, frameworks, or textbooks. I use a more dynamic and systemic view of the science curriculum, one that includes science content, the actions and behaviors of teachers and learners, and the various technologies of teaching.

Other features of a science curriculum include structure, function, and feedback. Structure consists of the relationships among concepts identified in materials, such as syllabi, frameworks, workbooks, laboratory manuals, videodisks, tests, software, and textbooks. In general, structure is usually thought of as curriculum materials. It sometimes is referred to as the intended curriculum (Glatthorn, 1987; Murnane and Raizen, 1988; Posner, 1992). The structured materials have been rationally thought out and planned by curriculum developers and likely express a particular emphasis (Roberts, 1995), scope, sequence, and organization.

Function consists of the many ways science teachers modify, adjust, and change the structured materials to accommodate classroom situations involving individuals and groups of learners. Function includes actions and behaviors of teachers and learners, for instance, questions, teacher-directed discussion, inquiry-oriented investigation, and instruction using educational technologies. This discussion elaborates the idea of actual or taught dimensions of the science curriculum (Glatthorn, 1987; Murnane and Raizen, 1988; Posner, 1992).

Feedback involves assessment of student attainment and the opportunities learners have to develop the understandings and abilities identified in the structured materials. Taking the end-of-lesson or course perspective, this is the achieved curriculum (Glatthorn, 1987; Murnane and Raizen, 1988; Posner, 1992). But I use *feedback* in the systemic sense. The feedback should serve to modify the structural and functional aspects of the curriculum. It certainly also should serve to identify the degree to which learners have achieved science knowledge, abilities, and understandings identified in standards.

The initial phase of the *National Science Education Standards* project used the term *curriculum*. The project was to develop standards for curriculum, teaching, and assessment (National Research Council, 1992). Early discussions convinced us that using the term *curriculum* would lead to the perception of a national curriculum for science—a view we did not believe in and one we did not wish to perpetuate. We changed from *curriculum* to *Content Standards* and found that many individuals confused content and curriculum anyway. Unfortunately, this confusion reflects poorly on the science education community because we demonstrate considerable confusion over seemingly elementary ideas about education.

As discussions continued, we also recognized the need for a perspective that was larger than the classroom orientation. The Program Standards and System Standards express this larger vision and some issues, such as financial support, equity, and a full K–12 perspective on school science programs. The Program Standards call for consistency, opportunities for all learners to achieve all the Content Standards, connections with mathematics, adequate resources, assurance of equity and access to opportunities to learn science content, and encouraging support for science teachers. Basically, the Program Standards recognize that achieving scientific literacy over the K–12 grades requires more than the sum of individual decisions about the science curriculum at respective grade levels. Further, there are some issues, such as financial resources and support for science teachers, that require decisions by school personnel not in classrooms.

Responsibilities of the Curriculum Developer

Regardless of their perspective—local, state, or national—science curriculum developers have responsibilities to several groups, including students, teachers, scientists and engineers, and parents and citizens in general.

Curriculum developers have a responsibility to provide all students with the best possible opportunities to learn. On the surface, this assertion may seem simple enough; it involves developmentally appropriate activities aligned with standards. On further analysis, however, one may recognize more elusive ideas, such as incorporating engaging and interesting experiences that have meaningful connections for learners, and activities that encourage those who often do not study and achieve in science, especially minority students. In working on these challenges, curriculum developers will find *Breaking the Barriers* (Clewell, Anderson, and Thorpe, 1992), *Contextual Factors in Education* (Cole and Griffin, 1987), and the work of Jeannie Oakes (1990a, 1990b) particularly insightful and helpful resources.

Curriculum developers should design materials that optimize teachers' knowledge and abilities. Curriculum materials should be understandable, manageable, and usable in the classroom. Additional teacher resources can assist continuing professional development through background readings in science and pedagogy, and through suggestions for additional experiences. Curriculum developers also should provide suggestions for teachers who wish to improve the science curriculum through adaptation. The availability and thoroughness of teacher resource materials is a critical test of locally developed programs, and most fail the test.

Curriculum developers have an obligation to represent science accurately and thoroughly. For example, the science curriculum should adequately represent the domains of science, and where appropriate, technology. Those who are developing curriculum have some responsibility to the major conceptual schemes that form the science disciplines. This responsibility is, however, not usually a problem. Rather, there is a problem of incorporating *too much* science content into a curriculum.

Science curriculum developers also have a responsibility to accurately represent science as a way of knowing and to appropriately defend against those who would introduce into the science curriculum topics that are nonscience. Creationism is a good example of what I mean. The science curriculum should help learners to understand science as a way of knowing that is based on empirical data, logical argument, and skeptical review.

In recent years, various groups have confronted curriculum developers, publishers, adoption committees, and science teachers with curriculum materials or topics that they think should be included in the science curriculum. Such groups claim the materials or topics are science, but on analysis they are not. Gerald Holton's book *Science and Anti-Science* (1993) and *Higher Superstition,* by Paul Gross and Norman Levitt (1994), address these issues. A specific example includes a situation that we continually encountered at BSCS: the proposal to include creationism in the science curriculum in addition to or instead of evolution. In my view, those who develop science curriculum demonstrate their responsibility to science when they provide materials demonstrating the integrity of science and support those who understand that nonscience positions have no place in school science programs.

Responsibility to the integrity of science presents problems to all of us in science education, but the issue is particularly troublesome for those developing, implementing, and teaching at the local level. School personnel are often ill prepared for political confrontations, and lack the understanding and support needed to defend the integrity of science. School personnel often take the easy way out of the conflict and yield to "equal time" arguments or avoid the word *evolution,* using instead the phrase "change over time." Such positions erode the integrity of science and leave learners with inadequate understandings about science disciplines, in this case, biology, and especially about the nature of science.

Science curriculum developers should provide meaningful connections between the concepts and processes of science and the personal and social dimensions of students' lives. I discussed these issues in *Reforming Science Education: Social Perspectives and Personal Reflections* (Bybee, 1994). For some time, we have provided learners with a view of science that was generally disconnected from personal issues and societal challenges. We are now being called to incorporate such perspectives and provide a context for science content. Although we have done a reasonable job of addressing the science goals of science education, specifically, the knowledge and processes, we are now challenged to improve the science curriculum by including the educational goals, namely, the personal and social contexts of science. As I said in an earlier chapter, the vision in the national standards and the framework for scientific literacy thoroughly address the larger contextual issues. In my view, these perspectives should be included in any science curriculum, and their inclusion is the responsibility of curriculum developers.

The Role and Importance of Professional Curriculum Developers

In a utopian society, all science teachers would have time, budget, expertise, and personnel to develop science curriculum for their unique teaching style, students, and community. To point out the obvious, science teachers do not work in a utopian society. They work in an educational system with severe constraints on time and budget, and variations in students' knowledge, abilities, and motivations. I fear that the burdens of standards-based reform will fall on science teachers and that they may succumb to the weight, bringing the potential of this reform to a halt.

Those who design and develop science curriculum provide a valuable resource for teachers, but like most resources, science curriculum materials must be adapted to meet unique human needs. The important point here is the adaptation of curriculum materials by science teachers. Often, teachers adopt and use science curriculum materials without adaptation. The process of adaptation by science teachers, whether the materials were developed at the national, state, or local levels, is one critical aspect of improving the science curriculum.

You might expect me, with a history as a curriculum developer, to defend the work of groups such as the BSCS, LHS, ACS, EDC, NSRC, TERC. I do not mind averring the importance of specialized groups for curriculum development. I am not the first to defend them (Karplus, 1971). Groups specializing in the development of science curriculum provide high-quality materials that meet various requirements for local and state curricula. Most science teachers, school districts, and state agencies do not have the technical capabilities, personnel, time, and money to develop science curricula and meet criteria of field testing in schools, coordinating content, incorporating sound pedagogy, and aligning assessment with standards, while also recognizing personal and social demands. The most important benefit of locally developed science curricula is ownership of the program and subsequent use by school personnel. Beyond this, very few locally developed materials have solved the problems of incorporating accurate science content, varied instructional approaches, and innovative assessment strategies into a complete science curriculum. Most of us did not build our own VCRs, computers, and cars. Local ownership can be achieved through professional development that focuses on the adaptation of materials originally developed outside the local district.

Having said this, I acknowledge that for many reasons science teachers, school districts, and state agencies will engage in development efforts as they align curriculum with science education standards. The discussion in the next section should help in such efforts.

Design Criteria for Improving the Science Curriculum

This section describes some requirements for contemporary science curriculum. I have found these criteria important in the design and development programs while at BSCS and consistent with many other programs. Individuals or groups can use these principles to evaluate current programs or as specifications for designing new curricula.

A Complete, Coherent, Consistent, and Coordinated Framework for Science Content

In several places, I have said that a science curriculum should provide all students with the opportunity to learn all content described in the national standards. This is the criterion of completeness. The framework for science should recognize the complete set of standards for science content. The science content also should be consistent with national, state, and local standards and benchmarks. Whether for lessons, units, or a complete elementary, middle, or high school program, the content should be well thought out, coordinated, and conceptually, procedurally, and contextually organized. That is, the roles of science concepts, inquiry, personal and social contexts, and the history and nature of science should be clear and explicit.

State and local curriculum frameworks provide a common reference for the development of coherent, consistent, and coordinated school science programs. Although curriculum frameworks have structure, they should allow for flexibility by those who have to accommodate different grade levels, science disciplines, and students. In recent years, curriculum frameworks, particularly at the state level, have shifted to align with national standards and to be used in the process of curriculum revision. Newer frameworks describe various components of the curriculum and serve to reduce fragmentation of different aspects of school science programs, for instance, content, teaching, and assessment (Curry and Temple, 1992).

A Balance of Knowledge, Understanding, and Abilities Associated with Different Dimensions of Scientific Literacy

Following is a list of the dimensions of scientific literacy described in Chapter 6. The science curriculum should represent a balance of opportunities for learners to develop functional, conceptual and procedural, and multidimensional scientific literacy. Traditionally, the science curriculum gave extensive emphasis to functional scientific literacy; contemporary criteria suggest greater balance, especially the inclusion of multidimensional scientific literacy.

Nominal Scientific Literacy

- Identifies terms, questions, as scientific.
- Demonstrates misconceptions.
- Has naive explanations.
- Shows minimal understanding.

Functional Scientific Literacy

- Uses scientific vocabulary.
- Defines terms correctly.
- Memorizes special responses.
- Understands only a specific need or activity.

Conceptual and Procedural Scientific Literacy

- Understands conceptual schemes of science.
- Understands procedural knowledge and skills of science.
- Understands relationships among parts and whole of science.
- Understands organizing principles, disciplines, and processes of science.

Multidimensional Scientific Literacy

- Understands the place of science among other disciplines.
- Knows the history of science.

- Knows the nature of science.

- Understands the interactions between science and society.

An Organized and Systematic Approach to Instruction

Most contemporary science curricula incorporate an instructional model. The instructional model should (1) provide for different forms of interaction among learners and between the teachers and learners; (2) allow for a variety of teaching strategies, such as inquiry-oriented investigations, cooperative groups, and use of technology; and (3) allow adequate time and opportunities for learners to formulate knowledge, skills, and attitudes. The following is the instructional model used in BSCS programs for elementary, middle, and high school programs designed after 1985. I describe this model in Chapter 8.

Engagement

This phase of the instructional model initiates the learning task. The activity should make connections between past and present learning experiences, anticipate activities, and focus students' thinking on the learning outcomes of current activities. The student should become mentally engaged in the concept, process, or skill to be explored.

Exploration

This phase of the teaching model provides students with a common base of experiences within which they identify and develop current concepts, processes, and skills. During this phase, students actively explore their environment or manipulate materials.

Explanation

This phase of the instructional model focuses students' attention on a particular aspect of their engagement and exploration experiences and provides opportunities for them to verbalize their conceptual understanding, or demonstrate their skills or behaviors. This phase also provides opportunities for teachers to introduce a formal label or definition for a concept, process, skill, or behavior.

Elaboration

This phase of the teaching model challenges and extends students' conceptual understanding and allows further opportunity for students to practice desired skills and behaviors. Through new experiences, the students develop deeper and broader understanding, more information, and adequate skills.

Evaluation

This phase of the teaching model encourages students to assess their understanding and abilities and provides opportunities for teachers to evaluate student progress toward achieving the educational objectives.

A Foundation of Psychological Principles Relative to Cognition, Motivation, Development, and Social Psychology

These psychological principles should be applied to the framework for content, teaching, and assessment. I recommend the *Learner-Centered Psychological Principles: Guidelines for School Redesign and Reform* (American Psychological Association, 1993). The twelve principles in summary are as follows:

1. The nature of the learning process. Learning is a natural process of pursuing personally meaningful goals, and it is active, volitional, and internally mediated; it is a process of discovering and constructing meaning from information and experience, filtered through the learner's unique perceptions, thoughts, and feelings.

2. The goals of the learning process. The learner seeks to create meaningful, coherent representations of knowledge regardless of the quantity and quality of data available.

3. The construction of knowledge. The learner links new information with existing and future-oriented knowledge in uniquely meaningful ways.

4. Higher-order thinking. Strategies for "thinking about thinking"—for overseeing and monitoring mental operations—facilitate creative and critical thinking and the development of expertise.

5. Motivational influences on learning. The depth and breadth of information processed, and what and how much is learned and remembered, are influenced by self-awareness and beliefs about personal control, competence, and ability; clarity and saliency of personal values, interests, and goals; personal expectations for success or failure; affect, emotion, and general states of mind; and the resulting motivation to learn.

6. Intrinsic motivation to learn. Individuals are naturally curious and enjoy learning, but intense negative cognitions and emotions (e.g., feeling insecure, worrying about failure, being self-conscious or shy, and fearing corporal punishment, ridicule, or stigmatizing labels) thwart this enthusiasm.

7. Characteristics of motivation-enhancing learning tasks. Curiosity, creativity, and higher-order thinking are stimulated by relevant, authentic learning tasks of optimal difficulty and novelty for each student.

8. Developmental constraints and opportunities. Individuals progress through stages of physical, intellectual, emotional, and social development that are a function of unique genetic and environmental factors.

9. Social and cultural diversity. Learning is facilitated by social interactions and communication with others in flexible, diverse (in age, culture, family background, etc.), and adaptive instructional settings.

10. Social acceptance, self-esteem, and learning. Learning and self-esteem are

heightened when individuals are in respectful and caring relationships with others who see their potential, genuinely appreciate their unique talents, and accept them as individuals.

11. Individual differences in learning. Although basic principles of learning, motivation, and effective instruction apply to all learners (regardless of ethnicity, race, gender, physical ability, religion, or socioeconomic status), learners have different capabilities and preferences for learning mode and strategies. These differences are a function of environment (what is learned and communicated in different cultures or other social groups) and heredity (what occurs naturally as a function of genes).

12. Cognitive filters. Personal beliefs, thoughts, and understandings resulting from prior learning and interpretations become the individual's basis for constructing reality and interpreting life experiences.

A Variety of Curriculum Emphases

Science curricula can, for example, emphasize conceptual schemes, inquiry, STS, and other orientations. No one curriculum emphasis is best for all learners. Probably a variety of emphases accommodate the interests, strengths, and demands of science content. Following are seven curriculum emphases described by Douglas Roberts (1994).

- everyday applications
- structure of science
- science/technology/decisions
- scientific processes
- solid foundation
- correct explanations
- self as explainer

Table 7.1 displays the curriculum emphases in the BSCS program *Middle School Science and Technology*.

Teaching Methods and Assessment Strategies Consistent with the Goal of Scientific Literacy

If the goals include inquiry and science in social perspectives, then it seems clear that inquiry-oriented teaching and social issues ought to be consistent approaches to teaching and assessment. The following are the organizing principles from the *National Science Education Standards* (National Research Council, 1996a) for teaching and assessment.

TABLE 7.1 *Curriculum Emphases in the BSCS Program* Middle School Science and Technology

Level A: Patterns of Change			
Unit	*1*	*2*	*3*	*4*
Curriculum Emphasis	Personal dimensions of science and technology	The nature of scientific explanations	Technological problem solving	Science and technology in society
Focus Question	How does my world change?	How do we explain patterns of change on the earth?	How do we adjust to patterns of change?	How can we change patterns?

Level B: Diversity and Limits			
Unit	*1*	*2*	*3*	*4*
Curriculum Emphasis	Personal dimensions of science and technology	Technological problem solving	The nature of scientific explanations	Science and technology in society
Focus Question	What is normal?	How does technology account for my limits?	Why are things different?	Why are we different?

Level C: Systems and Change			
Unit	*1*	*2*	*3*	*4*
Subtheme	Systems in balance	Change through time	Energy in systems	Populations
Curriculum Emphasis	Personal dimensions of science and technology	The nature of scientific explanations	Technological problem solving	Science and technology in society
Focus Question	How much can things change and still stay the same?	How do things change through time?	How can we improve our use of energy?	What are the limits to growth?

Teaching

- Plan an inquiry-based science program.
- Guide and facilitate learning.
- Assess teaching and learning.
- Design and manage educational environments.

- Develop and maintain a community of learners.

- Participate in school science program and policy decisions.

Assessment

- Assessments are consistent with decisions they are designed to inform.

- Achievement and opportunity to learn science must be assessed.

- Technical quality of the data collected is matched to decisions and actions taken on the basis of their interpretation.

- Assessment practices must be fair.

- Inferences made from assessments about student achievement and opportunity to learn must be sound.

Provision for the Professional Development of Science Teachers Who Will Have to Implement the Program

Research indicates that professional development is critical to the implementation of science curriculum (Hall and Hord, 1987; Fullan, 1982; Little, 1993; Loucks-Horsley and Stiegelbauer, 1991). This criterion is often ignored but absolutely essential. Following are the principles of professional development from the *National Science Education Standards*.

Professional Development

- Learn science content.

- Learn science teaching.

- Become life-long learner.

- Follow criteria for teacher development.

A Thorough Field Test and Review for Scientific Accuracy and Pedagogic Quality

One important legacy of the 1960s curriculum reform is the field testing of materials in a variety of science classrooms. For major BSCS programs, for example, we included two complete years of field testing and review. Without exception, field testing and reviewing a program identifies problems that developers did not recognize and fine-tunes the materials to the varied needs of teachers and learners and the requirements of schools. I also point out the importance of having scientists and engineers review science content for accuracy. Developers can miss subtleties of science or engineering concepts, inquiry, and design. Further, educators who review materials can provide valuable insights about teaching and assessment that help developers improve materials and enhance student learning.

★   ★   ★

These specifications provide a set of criteria that will contribute to the improvement of the science curriculum. Clearly, no one curriculum incorporates all specifications. There are always trade-offs when developing or adapting science curriculum.

Some Questions About Improving the Science Curriculum

Discussions of the science curriculum inevitably emerge within the larger context of educational reform and the process of establishing standards. During the process of developing national standards for science education, there came a time, early in 1995, when the majority of questions and discussions turned from the standards themselves to the curriculum questions. Most of the individuals and groups who had responsibilities for implementing the standards then began to shift attention to broad curriculum issues. In particular, we heard the curriculum questions from state and district science supervisors, professional curriculum developers, principal investigators of NSF State, Urban, and Local Initiatives, adoption committees for local school districts, and college personnel responsible for preservice and professional development of science teachers (see, for example, Bybee and McInerney, 1995). Basically, these workers accepted the national standards and were ready to initiate the difficult task of implementing the *Standards*. Predictably, the shift in attention from standards to the curriculum brought into focus a number of questions, some of which I address in this section.

What Are the Connections Among Scientific Literacy, National Standards, and the Science Curriculum?

The goal of achieving scientific literacy is broad and abstract. The good news is that the broad and abstract nature of the goal—scientific literacy—allows widespread agreement and a unifying idea for diverse groups within the science education community. For instance, science teachers, administrators, parents, curriculum developers, teacher educators, scientists, and policy makers can all support the idea of achieving scientific literacy. The bad news is that these diverse groups within the science education community have different perspectives on what scientific literacy is and how to achieve it. In general, the step from the broad goal of achieving scientific literacy to actual changes in school science programs and classroom practices is too large for many educators to make. Most individuals need intermediate statements that are more specific and concrete to guide their translation of the broad goal into the specific context of their work as science teachers, curriculum developers, science supervisors, and teacher educators.

Connections between the goals of achieving scientific literacy and improving the science curriculum can be found in documents that provide concrete guidance and information about what scientific literacy means. *National Science Education Standards* (National Research Council, 1996a) and *Benchmarks for Science Literacy*

(American Association for the Advancement of Science, 1993) are two such examples. The *Standards* can serve a variety of needs for those who have to do something about achieving scientific literacy. Notice that standards shift the perspective on scientific literacy from vision to concrete outcomes but still allow for local flexibility and individual variation in revising the science curriculum. Saying this nation is going to achieve higher levels of scientific literacy is like saying we are going to set the thermostat higher. Standards help sort out what *higher* means in Wisconsin, California, Texas, Maine, and Colorado, and more specifically, in local school districts in these states. Being relatively specific, yet general enough for use in state frameworks, teacher education, and curriculum improvement, is a great virtue of standards, whether national, state, or local.

Standards must be translated; they must guide the development of curriculum, instruction, and assessment. Of course, assessment can be done at the culmination of formal education, and at intermediate points during the learner's school career, such as grades 4 and 8. Thus, this point about assessment underscores the fact that standards act as criteria of attainment as well as being guidelines for teaching and learning. Standards include both inputs, such as curriculum, and outputs, such as student achievement. In the end, it is what students know and can do that will count as a measure of our progress toward achieving scientific literacy.

What Are the Implications for the Science Curriculum of the Framework for Scientific Literacy?

The framework and operational definition of scientific literacy proposed in Chapter 4 have significant implications for the science curriculum. Recall that the framework incorporated all the standards. Since the 1970s we have seen a steady change in the science curriculum toward an overemphasis on a single dimension of scientific literacy, namely, the functional. Major evaluations such as *Project Synthesis* (Harms and Yager, 1981) reported the textbook to be the curriculum, and most textbooks present an encyclopedic approach to science.

Improving the science curriculum so that it includes different dimensions of scientific literacy certainly requires a new view of the scope, sequence, experiences, and contexts of science in school programs. Table 7.2 uses a broad conception of the curriculum to identify some changes implied by the comprehensive view of scientific literacy.

What Does It Mean to Say a Curriculum Aligns with Standards?

This question addresses several important issues associated with the implementation of standards. The curriculum being alluded to is usually the current program in a school district or extant materials such as a BSCS program or other commercially available textbooks. To be clear at the beginning of this discussion, such a question reveals several fallacies. First, no science curriculum will align perfectly with standards, be they national, state, or local. The only thing that will match perfectly are the standards themselves. Although this may seem obvious, some

TABLE 7.2 *Implications of Framework for Scientific Literacy for Improving the Science Curriculum*

Elements of Scientific Literacy	Curriculum Materials	Instructional Methods	Assessment Strategies
Learners demonstrating *functional scientific literacy* respond adequately and appropriately to vocabulary and technical words associated with science and technology. Learners meet minimum standards of literacy. They can read and write passages with simple scientific and technological vocabulary. Learners may associate vocabulary with larger conceptual schemes, but they demonstrate little knowledge of scientific concepts, principles, laws, theories, and the fundamental procedures and processes of scientific inquiry.	Vocabulary introduced on a need-to-use basis. Ideas and skills experienced before formal introduction.	Introduction of vocabulary in course of instructional sequences.	Vocabulary reviewed in context of assessments such as portfolios and performances or investigations. Assessments include knowledge of technical terms and vocabulary in science.
Conceptual and procedural literacy occurs when learners demonstrate an understanding of major concepts and processes of traditional domains such as physical, life, and earth sciences. Learners can identify the way new explanations and inventions develop through the processes of inquiry and design. Learners understand the structure of science disciplines and the procedures for developing new knowledge and techniques.	Concepts and processes of standard made explicit. Inquiry emphasizes at least one full inquiry a year. All content standards included in school science program.	Systematic and sequential phases of instruction. Time and multiple opportunities to learn. Individual, small group, and large groups.	Achievement based on authentic performance. Some assessment determination in context of instructional sequence. Assessment includes knowledge of science concepts and procedures.

TABLE 7.2 *Implications of Framework for Scientific Literacy for Improving the Science Curriculum (continued)*

Elements of Scientific Literacy	Curriculum Materials	Instructional Methods	Assessment Strategies
Multidimensional literacy has two components, one that integrates various aspects of scientific concepts and procedures and one that enlarges understandings through study of science in contexts of other disciplines, society, and history. *Integral literacy* consists of understanding the essential conceptual structures of science plus features that make that understanding of the disciplines more complete. In *contextual literacy*, learners understand the relationship of disciplines to the whole of science and technology and to various personal issues and societal challenges.	Historical examples used to show development of science and procedures.	Sequence of lessons includes connections to other domains and topics.	Assessment includes science and technology; science in personal and social perspectives, and history and nature of science.
	Investigations include science- and technology-related social issues.		Assessment includes understanding of science concepts and procedures in personal and social contexts.
	Connections to personal aspects of science.		
	Variety of curriculum emphases with the science curriculum.		
	Opportunities to learn about interactions of science with other disciplines and with society.		

individuals still want a complete alignment or they decide that the curriculum under review is inadequate or incomplete. Second, curriculum materials should not be assessed using a single criterion, even if this is standards. Especially in programs developed from a national perspective, decisions about the organization and presentation of materials must accommodate multiple criteria, such as manageability and usability by teachers and learners, safety, and adoption requirements, such as legal compliance. Third, it is difficult to use science content as the principal criterion for judging the alignment of a science curriculum with standards. All publishers claim that their textbooks and materials align with national standards.

Some ask if a certain curriculum *meets* the standards. This reveals a lack of understanding about the role of standards as outcomes. "Meeting the standards" is correctly interpreted as students achieving the knowledge, abilities, and understandings defined in the standards.

Having addressed a number of issues associated with the alignment between a science curriculum and science education standards, individuals will, with good intentions, seek to improve the science curriculum through alignment with standards. In the discussion that follows, the term *indicators* describes features or attributes that show the degree of alignment. The discussion generally progresses from the simplest to more complex attributes of a science curriculum.

In the simplest analysis, one can examine instructional materials to see if lessons or units are aligned with actual content domains, conceptual organizers, and fundamental understandings in the *Standards*. This approach, however, only gives an estimate of alignment. One must assume that science content from all standards will not be in the curriculum under review.

Probably a better way to analyze alignment is to identify what is or is not evident in comparing the curriculum materials and the standards. What about the science content that is in the standards but not in the curriculum? For example, suppose the analysis reveals considerable alignment with subject matter in the physical and life sciences but omissions concerning standards on the history and nature of science. What about the omissions? This example points to the importance of a curriculum framework at the state and local levels and the need to design total school science programs that are vertically integrated. Such a framework gives a larger perspective on the science education program and thus identifies the opportunities to address omissions in other parts of the school science program.

Taking the analysis to a deeper level, we point out that activities within the curriculum should be examined to determine how closely and explicitly they align with (1) the knowledge, abilities, and understandings in the standards; (2) the time and opportunity to achieve the knowledge, skills, and understandings in the standards; and (3) the orientation and emphasis of activities in the teaching standards. Also, (4) does assessment align with content, and (5) does the curriculum identify what counts as evidence of achievement for the learner and teacher so they know if and when the desired level of achievement has been attained?

Let me clarify what I mean by the explicit alignment of the activities with standards. I think it is essential to explicitly teach a concept or skill. If you want students to know that ecosystems have limiting factors and that using oceans as a commons can result in ecological tragedy, then you have to teach that directly. The artistic metaphor of foreground and background applies. Such ideas about ecosystems may be in the background of various opportunities to learn, but at some point, they have to be the explicit point of a lesson or activity. There are too many examples of materials that include topics aligned with the content standards but do not address the fundamental understandings and abilities. Honorable mention is not enough to count as alignment of a curriculum with standards. A critical test for this indicator would be to select a number of fundamental understandings from the standards and try to identify the activities that explicitly develop the understandings. For instance, when examining a science curriculum for middle school, you should be able to identify activities that develop understandings, such as the following:

- Substances have characteristic properties, such as density, boiling point, and solubility, which are independent of the amount of the sample.

- An organism's behavior has evolved through adaptations to its environment.

- Global patterns of atmospheric movement influences local weather.

Although these examples come from the subject matter standards of physical, life, and earth science, you should not ignore the other content standards. Also, time and opportunity to learn are critical indicators. Contemporary emphasis on constructive approaches to learning suggests that single lessons are probably inadequate approaches to learning because they do not provide learners with the time to engage in an idea, explore alternatives, discover different explanations, and construct more meaningful and scientifically accurate explanations. This process requires more time and multiple opportunities to develop understanding. The use of instructional models in curriculum materials, or the potential of nonlinear approaches of educational technologies, usually accommodate the indicators of time and multiple opportunities to learn. A fairly clear indication of a program that does not meet this indicator is one that has 180 different activities for 180 days and more than 180 vocabulary words.

Another indicator combines opportunities to learn and instruction. Suppose a science textbook gives substantial emphasis to the memorization of vocabulary and the fundamental understandings. What is wrong with this alignment? The indicators have to do with the standards' emphasis on the opportunities for students to engage in investigations and the use of inquiry as a teaching method.

Finally, two indicators center on assessment. The first asks whether the curriculum includes assessment consistent with the content and assessment standards. The second requires a close look at assessment in curriculum materials to see if it defines the acceptable levels of achievement. Does the science curriculum provide adequate information about acceptable goals of achievement and what students say, write, and communicate that indicates they have achieved the knowledge, skills, and understandings of the standards? Ask questions about the form of assessment and its consistency with the goal of scientific literacy. For example, the goal of scientific literacy in the *National Science Education Standards* does not center on vocabulary. Although not entirely excluding technical words, the emphasis is much more conceptual, procedural, and multidimensional so an assessment that focuses on vocabulary defines a type of scientific literacy that varies from an alignment with national standards. Following is a summary of the indicators of curriculum alignment with standards:

Indicators	Levels of Analysis
Identification of science content	Do topics of curriculum match the conceptual organizers of standards? Are standards explicitly represented in the curriculum framework?
Explicit connections with fundamental abilities and understandings	Do activities include the fundamental understandings and abilities of the standards? Do activities include all the fundamental understandings and abilities for a particular standard? Are foreground and background emphasis of concepts and abilities made explicit? Are connections made with other topics, concepts, and procedures?
Time and opportunities to learn	Does instruction include several activities on a topic? Do students experience concepts before terms are introduced? Do students apply concepts and procedures in different contexts?
Appropriate and varied instruction	Are different methods of instruction used? Are students engaged in investigations with an inquiry orientation? Are teachers informed of possible misconceptions and how to encourage conceptual change?
Appropriate and varied assessment	Are opportunities provided for teachers to identify what students know and can do? Are assessment strategies consistent with philosophy or pedagogy? Are assessments comprehensive, coherent, and focused on the understandings of science content and abilities to use procedures?
Clarification of achievement	Are there examples of acceptable and unacceptable levels of achievement?

In final analysis of the alignment of a curriculum with standards, it is probably more productive to ask a different question: What do we have to do to align our curriculum with standards? Once you establish a curriculum framework that provides a comprehensive view of a proposed school program, you will be in a much better position to compare the extant curriculum, review a curriculum proposed for adoption, or produce your own curriculum. At this point, my recommendation should be clear—for most school districts, time and money are best spent on improving the science curriculum through professional development that emphasizes adaptation of materials.

What Are the Requirements for a Curriculum That Enhances Student Learning and Achievement of Scientific Literacy?

To a large degree, I answered this question in the section on "Design Criteria for Improving the Science Curriculum." But it may be important to restate and elaborate several ideas. One requirement for improving the science curriculum is shifting perceptions from science lessons to student learning. This shift may be evident to most individuals, but some still view teaching science lessons as equivalent to meeting a standard. If a science curriculum includes a unit on environmental quality (grades 9–12), then that is considered sufficient. The shift in perception suggests that a unit on environmental quality is probably important, but the standard requires that students *develop understandings,* such as that natural ecosystems provide an array of processes that affect humans; materials from human societies disturb both physical and chemical cycles of the earth; and many factors like population growth, resource use, population distribution, economics, and technology influence environmental quality. Such understandings may or may not be developed in a unit specifically on environmental quality.

A second requirement involves expanding perceptions of learning experiences. The artistic metaphor of foreground and background clarifies this idea. Most artistic renderings have something in the foreground and other things in the background. The trained observer understands that the background is an essential part of the picture. With regard to the science curriculum, a prominent misunderstanding suggests that an opportunity to learn a fundamental understanding would include a single experience (lesson). Such perceptions are exemplified when one hears such statements as, "There are *x* number of fundamental understandings and abilities for grades 5–8; that means I have to devote *x* number of time slots per day per concept and ability over the four years." On the face of it, this analysis conflicts with our views of teaching and learning, yet such analyses are common. Views such as this betray a perception that learning only occurs with a single experience. Although fundamental concepts and abilities have to be made explicit to learners, I also have suggested that concepts and abilities develop through multiple experiences in different contexts. Concepts and abilities may be in the background of student experience. It is up to teachers to pull ideas to the foreground by pointing out the connections and relationships among concepts.

Sometimes our understandings of science and beliefs about education detract from experiences that enhance student learning. I recently listened to several individuals describe a unit they had designed on an endangered species. Their initial view was that this unit met the state's ecology standard. They told of field trips the students would take, mapping skills they would develop, birds they would identify, and habitats they would observe. I asked where the students would learn about limiting factors in ecosystems, competition for resources, and what causes a species to become endangered? The teachers looked puzzled, became defensive, and responded, "All the concepts [in the standards] are there. The students will have wonderful experiences, love the activities, and have fun and learn a lot." I asked, "Where do you specifically address the concepts outlined in the standards, and how will you know what the students learned?" This is an example of standards getting honorable mention, or students having opportunities to learn because they were "in the territory." It is no longer enough for students to just have "wonderful experiences"; they have to learn science concepts and develop cognitive abilities as well. Certainly, any instructional sequence should have engaging and interesting experiences, but these should result in students' developing understandings and abilities.

What Is Involved in Translating Standards to a Science Curriculum?

Translating standards to a science curriculum involves developing a new curriculum or adapting an extant curriculum. This section addresses the development of new materials. Obviously, I am not going to design an entire science curriculum here; rather, I will limit the discussion to activities that serve as examples of the process.

The example outlined in the following list begins with the standard Science as Inquiry for grades 9–12. After reviewing details of the standard and deciding that developing abilities of scientific inquiry would be the object of the activities, it is important to review the details, the fundamental abilities of that standard. This leads to ideas about activities and to the selection of activities that would best develop the abilities that will be emphasized. At this point, you should decide on the various aspects of the activity and review consistency of the activity with teaching standards. Finally, you should determine the indicators of achievement. Although the list is sequential, in reality several of these steps probably will occur simultaneously.

Inquiry Standard

- As a result of activities in grades 9–12, all students should develop the abilities of scientific inquiry and understandings about scientific inquiry.

Review Fundamental Abilities

- Identify questions and concepts that guide scientific investigations.

- Design and conduct full investigations.

- Use technology to improve investigations and communications.

- Formulate and revise scientific models using logic and evidence.

- Recognize and analyze alternative explanations and models.

- Communicate and defend a scientific argument.

Identify Appropriate Opportunities to Learn

- Students review current report and identify questions and concepts.

- Students conduct a full investigation.

- Students argue data and formulate and communicate explanations.

Decide on Activity and Design Instructional Sequence

- Students work in groups of three.

- Each group has similar data on the use of insecticide to eradicate insects.

- Students formulate mechanism that explains the fact that all insects are not eradicated and the remaining population continues to grow.

- Students communicate their explanations.

- Other student groups analyze explanations and propose alternative explanations.

Review Teaching Standards

- There is time and opportunity to develop abilities.

- Activity and instruction align directly with standard.

- Activity is developmentally appropriate.

- Activities use "community of learners" approach.

Determine Indicators of Achievement

- Students formulated explanation that accounted for data.

- Students used scientific knowledge.

- Students demonstrated logic, knowledge, and reasoning to construct explanation.

- Students critically analyzed other arguments.

- Students communicated accurately.

- Students communicated effectively by using charts and diagrams.

- Students respond to critical questions by using scientific knowledge, data, and reasoning.

The more likely approach to translating standards to a science curriculum begins with an idea for an activity or a unit. I briefly describe here the design of a unit based on a classic article from the literature of human ecology: Garrett Hardin's "The Tragedy of the Commons" (1968). Discussing the design of this proposed program provides the opportunity to examine several issues about curriculum design and how standards are incorporated into new materials. This example also exemplifies an interdisciplinary approach to the science curriculum. First, I provide some background on the problem and then discuss the educational remedy of improving the science curriculum.

In recent decades, issues related to environment, resources, and population have developed from local to global problems. Twenty years ago, some forests in Germany were believed to be damaged by acid rain; now acid precipitation is widespread in Europe and North America. In the 1960s, ozone depletion was virtually not an issue; now there is continuing scientific research on the causes and effects of the progressive depletion of the ozone layer. In the past, extinction of plants and animals was an academic concern; now biodiversity and preservation of species are part of policy debates. Scientists and others understand that population growth contributes to the aforementioned problems (Brown et al., 1993).

Many reports describe the scientific, economic, policy, and ethical aspects of environmental and population problems, and these same reports inevitably identify education as an essential remedy for the problems (Sitarz, 1993; Silver and DeFries, 1990). Few scientists or educators, however, have discussed in detail the educational implications of these issues, much less answered the questions of how to design materials that appropriately and adequately present the interrelated scientific, political, and ethical aspects of the problems.

However, one article that presents an interdisciplinary analysis of our contemporary problems and provides a context for educational programs is Hardin's "The Tragedy of the Commons." In December 1968, *Science* published this article. Now, that interdisciplinary statement about the impact of human population on resources, the environment, and other humans is a classic analysis in environmental literature. Hardin presents several important insights about humans and their environments that connect to the *Standards*. Some of his insights are that many problems do not have technological solutions; many individual decisions are inexorably related to social conditions; there are natural limits to the use of natural resources; and ethics, economics, politics, and science are interrelated in fundamental ways.

The *Standards* provide a foundation for redress through the vehicle of curriculum materials. Is there a relationship between the science and societal issues described in the prior paragraphs and the *Standards?* The answer is yes. The *Standards* describe standards for life science and for science in personal and social perspectives. At grades 9–12, the standards in life science include Interdependence of Organisms, which is directly related to ecology and the aforementioned issues. Of more direct importance for this discussion, the topic Tragedy of the Commons also requires an interdisciplinary perspective and thus has connections to other standards on Science as Inquiry and Science in Personal and Social Perspectives, which include major conceptual organizers for population growth, natural resources, environmental quality, natural and human-induced hazards, and science and technology in local, national, and global challenges.

The proposed program "The Tragedy of the Commons: Understanding Science in Societal Challenges" provides an integrated approach to help learners develop the following major goals related to scientific literacy: (1) an understanding of the basic scientific principles related to the "tragedy of the commons," (2) development of skills for critical thinking and ethical analysis, and (3) recognition of the role of science in personal and social perspectives. The following lists identify some fundamental understandings, abilities, and concepts in the *National Science Education Standards* with which the proposed program "The Tragedy of the Commons" aligns. I have included all the fundamental concepts and abilities for the respective standards. Those fundamental understandings printed in italics serve as the primary objectives for the program. They are in the foreground.

NSES Fundamental Understandings from the Life Sciences: Grades 9–12

Biosphere and Interdependence

- Physical elements, such as water and minerals, cycle among the living and nonliving components of the biosphere. Considering the biosphere as a whole, the cycles include gaseous types with a large reservoir in the atmosphere and sedimentary types within reservoirs in the earth's crust.

- Energy flows through ecosystems. Ecosystems represent a portion of the biosphere, and energy flows through ecosystems in one direction, from photosynthesis to herbivores to carnivores and decomposers. Because about 90 percent of the stored energy is lost at each step, ecosystems comprise a limited number of stages for the flow of energy.

- The interrelationships and interdependencies between organisms that have evolved in ecosystems reflect both cooperation and competition. These interrelationships and interdependencies make the functioning of the ecosystem possible. Ecosystems are often stable for hundreds or thousands of years, but natural disturbances or changes in species composition often lead to changes. Changes in ecosystems can lead to new selective processes.

- Living organisms have the capacity to produce populations of infinite size, but environments and resources are finite. This fundamental tension has both short- and long-term influences on the interaction and interdependence of organisms and populations.

- *Human beings live within the world's ecosystems.* Increasingly, humans modify ecosystems as a result of population growth, technology, and consumption. Human destruction of habitats through direct harvesting, pollution, atmospheric changes, and other factors is threatening global stability, and if not addressed, ecosystems will be irreversibly damaged.

Science as Inquiry

- Identify questions and concepts that guide scientific investigations. Students should formulate a testable hypothesis and demonstrate the logical connections between the scientific concepts guiding a hypothesis and the design of an experiment. They should demonstrate procedures, a knowledge base, and conceptual understanding of scientific investigations.

- Design and conduct scientific investigations. Designing and conducting a scientific investigation requires introduction to conceptual areas of investigation, proper equipment, safety precautions, assistance with methodological problems, recommendations for use of technologies, clarification of ideas that guide the inquiry, and scientific knowledge obtained from sources other than the actual investigation. The investigation may also include such abilities as identification and clarification of the question, method, controls, and variables, the organization and display of data, the revision of methods and explanations, and the public presentation of the results and the critical response from peers. Regardless of the scientific investigations and procedures, they must use evidence, apply logic, and construct an argument for their proposed explanation.

- Use technology to improve investigations and communications. Students' ability to use a variety of technologies, such as hand tools, measuring instruments, and calculators, should be an integral component of scientific investigations. The use of computers for the collection, analysis, and display of data is also a part of this standard.

- *Formulate and revise scientific explanations and models using logic and evidence.* Student inquiries should culminate in formulating an explanation or model. In the process of answering the questions, the students should engage in discussions and arguments that result in the revision of their explanations. These discussions should be based on scientific knowledge, the use of logic, and evidence from their investigation.

- *Recognize and analyze alternative explanations and models.* This standard emphasizes the critical abilities of analyzing an argument by reviewing current scientific understanding, weighing the evidence, and examining the logic thus revealing which explanations and models are better and showing that although there may be several plausible explanations, they do not all have equal weight. Students should appeal to criteria for scientific explanations in order to determine which explanations are the best.

- *Communicate and defend a scientific argument.* Students in school science programs should develop the abilities associated with accurate and effective communication including writing and following procedures, expressing concepts, reviewing information, summarizing data, using language appropriately, developing diagrams and charts, explaining statistical analysis, speaking clearly and logically, constructing a reasoned argument, and responding to critical comments through the use of current data, past scientific knowledge, and present reasoning.

Fundamental Concepts from Science in Personal and Social Perspectives: Grades 9–12

Population Growth

- *Populations grow or decline through the combined effects of births and deaths, and in countries through emigration and immigration.* Populations, and other things such as resource use and environmental pollution, can increase through linear or exponential growth.

- Various factors influence birth rates and fertility rates, such as average levels of affluence and education, importance of children in the labor force, education and employment of women, infant mortality rates, costs of raising children, availability and reliability of birth control methods, religious beliefs and cultural norms that influence personal decisions about family size.

- *Populations can reach the limits to growth.* Carrying capacity is the maximum number of individuals that can be supported in a given environment. The fundamental understanding is not availability of space, it is the number of people in relation to resources and the capacity of earth systems to support human beings.

Science and Technology in Local, National, and Global Challenges

- Science and technology are essential social enterprises, but alone they can only indicate what can happen, not what should happen. The latter involves human decisions about the use of knowledge.

- *Understanding basic concepts and principles of science and technology should precede active debate about the economics, policies, politics, and ethics of various science- and technology-related challenges.* But, understanding science alone will not resolve local, national, or global challenges.

- Progress in science and technology can relate to social issues and challenges. Funding priorities and health problems serve as examples of ways that social issues influence science and technology.

- Individuals and society must decide on proposals involving new research and technologies. Decisions involve assessment of alternatives, risks, costs, and benefits and consideration of who benefits and who suffers, who pays and gains, and what are the risks and who bears them? Students should understand the appropriateness and value of basic questions: What can happen? What are the odds? and How do scientists and engineers know what will happen?

- *Humans have a major effect on other species.* The influence of humans on other organisms occurs through ways, such as land use—decreasing space available to other species, and pollution—changing the chemical composition of air, soil, and water.

Environmental Quality

- *Natural ecosystems provide an array of basic processes that affect humans.* Those processes include maintenance of the quality of the atmosphere, generation of soils, control of the hydrologic cycle, disposal of wastes, and recycling of nutrients. Humans are changing many of these basic processes and the changes may be detrimental to humans.

- Materials from human societies disturb both physical and chemical cycles of the earth.

- Many factors influence environmental quality. Factors that students might investigate include population growth, resource use, population distribution, overconsumption, the capacity of technology to solve problems, poverty, over-simplified views of earth systems, the role of economic, political, and religious views, and different ways humans view the earth.

Natural Resources

- *Human populations use resources in the environment in order to maintain and improve their existence.* Natural resources have been and will continue to be exploited to maintain human populations.

- *The earth does not have infinite resources, and increasing human production and consumption places severe stress on the natural processes that renew some resources and depletes those resources that cannot be renewed.*

- *Humans use many natural systems as resources.* Natural systems do have the capacity to reuse waste, but that capacity is limited. Changing natural systems can exceed the limits of organisms to adapt naturally or humans to adapt technologically.

In addition, the program "The Tragedy of the Commons" includes objects associated with the standards on teaching. It will help teachers (1) decrease their perception of science as a body of discipline-based information and increase their understanding of science as an integrated way of explaining natural phenomena and applying those explanations to analysis of societal challenges, (2) decrease their use of lecture/discussion and increase their use of inquiry and interactive instructional strategies, and (3) decrease their use of traditional tests and increase their use of alternative assessments.

The primary audience for the proposed program "Tragedy of the Commons" is students studying science in grades 9–12. These students may be in tenth-grade biology courses, or courses such as those proposed by the NSTA project *Scope, Sequence, and Coordination of Secondary School Science* (National Science Teachers Association, 1992). In addition, the materials are designed for use in other courses, such as second-level biology, STS, integrated sciences, environmental sciences, intradisciplinary science, and social studies. The program's modular design and use of interactive technologies should provide the flexibility required for implementation in a wide range of school programs.

The following list summarizes a number of specifications of the proposed program "Tragedy of the Commons." Developing a list of specifications helps identify the qualities of the program and the role of various components, and answers questions about the number of activities, use of the program, background information, and alignment with standards and benchmarks. Most likely, school personnel will not develop programs this complex. However, the general process of beginning with an idea for instructional materials and making connections to standards remains essentially the same.

Specifications for Instructional Materials in the Proposed Program "Tragedy of the Commons"

- Teacher's guide—approximately 200 pages.

- One video—combination motion and still images. Video will include background information; interviews with scientists, politicians, economists, and ethicists; databases; and models.

- Ten to twelve teaching activities (activities may have several lessons). Video will be integral to work on the activities. An instructional model will be the basis of the instructional approach. The model, known as the 5 Es, has the following phases of instruction: engage, explore, explain, elaborate, and evaluate.

- Topics on environment, resources, populations (including opposition to Hardin's perspectives). The program will include other "commons" problems such as health care, the U.S. budget, school cafeterias, national parks, and fishing off the New England coast. In general, the activities will enlarge student perspectives from local to global.

- Activities may be used in discipline-specific classes (biology, earth science) or interdisciplinary classes (environmental studies, STS) and may support mathematics through use of statistics, data analysis and presentation, and study of doubling time and exponential growth.

- Use in grades 9–12. In some cases, these materials will be applicable in college courses on ecology and environmental sciences and education courses for science teachers.

- Align with *National Science Education Standards* on Life Science, Inquiry, and Science in Personal and Societal Perspectives.

What Is It Like to Adapt a Curriculum to Align with Standards?

In short, the task is large and it will take slow and steady work over a considerable period of time. Few locally developed science programs meet the criteria discussed in this chapter. In my view, *adapt* is the operative word, and in most cases, adapting will be preceded by *adopting*. That is, most school districts should adopt curriculum materials and then adapt them through professional development.

If science teachers or local school districts are adapting their current materials to address the *Standards,* then the first issue is, How close are the extant units and lessons to specific standards? If the materials generally have features consistent with the philosophy, pedagogy, and content of standards, then adaptation may be fairly easy. If, on the other hand, extant materials show considerable variation from that philosophy, pedagogy, and content, the adaptation may simply be too much and school personnel should consider a science curriculum that most closely aligns with the local and state requirements, national standards, and the interests and talents of science teachers.

Conclusion

This chapter focused on the improvement of the science curriculum. The view of the science curriculum moves beyond the textbook as curriculum or hundreds of hands-on activities. The science curriculum is a designed set of relationships among content, teaching, context, and assessment. Regardless of your approach—revising the old or adopting the new—improving the science curriculum has costs, benefits, and trade-offs. If we continue to improve the science curriculum and maintain the focus on a standards-based systemic reform, then we should begin to realize our goal of scientific and technologic literacy for all students.

8

Improving Instruction

The heart of this chapter consists of an instructional model that should help science teachers improve their instructional practices. I am convinced that the ultimate reform of science education will only occur at the level of science classrooms. Recommending an instructional model touches the heart of teaching—the most practical level at which educational reform occurs.

The instructional model used extensively in several BSCS programs, commonly referred to as the 5 Es, consists of the following phases: engagement, exploration, explanation, elaboration, and evaluation. Each phase has a specific function and contributes to the teacher's coherent instruction and the students' constructing a better understanding of scientific and technological knowledge, attitudes, and skills. The model can be used to help frame the sequence and organization of programs, units, and lessons. Once internalized, it also can inform the many instantaneous decisions science teachers must make in classroom situations.

Constructivism is the theoretical foundation for this instructional model. The constructivist view assumes a dynamic and interactionist conception of human learning. Students bring to a learning experience their current explanations, attitudes, and skills. Through meaningful interactions between themselves and their environment, which includes other students and teachers, they redefine, replace, and reorganize their initial explanations, attitudes, and skills. In developing the instructional model, I assumed that the constructive process could be assisted by sequences of experiences designed to challenge current conceptions, attitudes, and skills, and provide time and opportunities for reconstruction to occur.

Historical Perspectives on Instructional Models

The 5 Es model has an orientation and elements identifiable in educational history. This section presents some historical antecedents of the contemporary 5 Es model. The section begins with a general characterization of learning theories and then describes instructional models proposed by individuals, such as Johann Friedrich Herbart, John Dewey, and Jean Piaget, all of which can be viewed as conceptual predecessors to the 5 Es model.

Views of Learning

Historically, educators have explained learning by classifying it into one of three broad categories: transmission, maturation, and construction (see Table 8.1). B. F. Skinner's theories serve as an example of the first category—transmission (Ferster and Skinner, 1957). Arnold Gesell's model exemplifies the second—maturation (Ilg and Ames, 1955). Many science educators would recognize Piaget's ideas as associated with the third category—construction (Piaget, 1975; Piaget and Inhelder, 1969). Piaget viewed development of the intellect as neither direct learning from the environment nor maturation. Rather, he proposed that learning consists of reorganization and reconstruction of psychological structures as a result of interactions between the individual and the environment. The following sections present several examples of individuals whose ideas contributed to contemporary perspectives of teaching and learning.

Herbart's Model

Johann Friedrich Herbart, a German philosopher, influenced American educational thought around the turn of the century. For Herbart, the primary purpose of education was development of character, and the process of developing character began with the student's interest. At a more specific level, Herbart considered concepts to be the fundamental building blocks of the mind, and the function of a concept constituted justification for including a concept in a course of study. In a contemporary sense, Herbart was interested in the creation and development of conceptual structures that would contribute to an individual's development of character. Herbart's philosophy contrasted with another model that proposed that the purpose of education was to exercise the mind (DeBoer, 1991; Tanner and Tanner, 1990).

TABLE 8.1 *Perspectives for Instructional Strategies*

Perspective	*Assumptions about Students as Learners*	*Views of Knowledge*	*Approaches to Teaching*
Transmission	Students must be filled with information and concepts.	Concepts are a copy of reality.	External to internal
Maturation	Students must be allowed to mature and develop.	Concepts emerge.	Internal to external
Construction	Students must be actively involved in learning.	Concepts are constructed.	Interaction between internal and external

Herbart proposed two principal ideas that he used as foundations for teaching: interest and conceptual understanding. The first principle of effective instruction consisted of the student's interest in the subject. Herbart suggested two types of interest, one based on direct experiences with the natural world and the second based on social interactions. Instruction can quite easily use the natural world and capitalize on the curiosity of students. In addition, teachers can introduce objects from the natural world and use them to help students accumulate a rich set of sense impressions. Herbart suggested the observation and collection of living organisms and the introduction of tools and machines. As teachers introduce organisms and objects, they should take into account and make connections to prior experiences (Herbart, 1901).

Herbart's model also suggested that teachers should recognize the social interests of children and interactions with other individuals. A thorough education incorporates social interactions and recognizes their contribution to learning. Thus, an instructional model should incorporate opportunities for social interaction among students, and between students and the teacher.

The second principle of Herbart's model resides in the association of the sense perceptions with generalizations or principles, or the formation of concepts. For Herbart, sense perceptions of objects, organisms, and events were essential, but in and of themselves they were not sufficient for the development of mind. A very important theme in Herbart's model is the coherence of ideas. That is, each new idea must be related to extant ideas. Said in contemporary terms, prior knowledge is the point of departure of instruction.

If we synthesized Herbart's ideas into an instructional model, as others have (see, for example, Compayre, 1907; DeBoer, 1991), we would begin with the current knowledge and experiences of the student and present ideas that easily related to those concepts the student already had. The introduction of new ideas that connected with extant ideas would slowly form concepts. According to Herbart (1901), the best pedagogy allowed students to discover the relationships among experiences. The teachers guided, questioned, and suggested through indirect methods. The next step in the instructional model involved direct instruction, where the teacher systematically explained ideas that the student could not be expected to independently discover. In the final step of Herbart's model, teachers asked students to demonstrate their conceptual understanding by applying the concepts to new situations. Students solved problems, wrote essays, and performed tasks that demonstrated their understanding of the concepts. Herbart's model is one of the first systematic approaches to teaching and has been used in various forms by educators for more than one hundred years (DeBoer, 1991). The following list summarizes Herbart's instructional model.

Preparation	The teacher brings prior experiences to students' awareness.
Presentation	The teacher introduces new experiences and makes connections to prior experiences.

Generalization	The teacher explains ideas and develops concepts for the students.
Application	The teacher provides experiences where the students demonstrate their understanding by applying concepts in new contexts.

Dewey's Model

John Dewey began his career as a science teacher. No doubt, the early influence of science explains the obvious connection between Dewey's conception of thinking and scientific inquiry. In *How We Think,* Dewey (1933 [1910]) outlined what he termed a complete act of thought and described what he maintained were indispensable traits of reflective thinking. The traits included (1) defining the problem, (2) noting conditions associated with the problem, (3) formulating a hypothesis for solving the problem, (4) elaborating the value of various solutions, and (5) testing the ideas to see which provided the best solution for the problem.

In *Democracy and Education,* Dewey (1916) further described the relationship between experience and thinking. He summarized the general features of the reflective experience:

> (i) perplexity, confusion, doubt, due to the fact that one is implicated in an incomplete situation whose full character is not yet determined; (ii) a conjectural anticipation—a tentative interpretation of the given elements, attributing to them a tendency to affect certain consequences; (iii) a careful survey (examination, inspection, exploration, analysis) of all attainable consideration which will define and clarify the problem in hand; (iv) a consequent elaboration of the tentative hypothesis to make it more precise and more consistent, because squaring with a wider range of facts; (v) taking one stand upon the project hypothesis as a plan of action which is applied to the existing state of affairs: doing something overtly to bring about the anticipated result, thereby testing the hypothesis. (p. 150)

Based on this quotation, it seems clear that Dewey implied an instructional approach that was based on experience and required reflective thinking. In contemporary terms, doing hands-on activities in science is not enough. Those experiences also must be minds-on. In Dewey's time, and in the contemporary reform, many individuals regard activity-based, hands-on, process-oriented programs as ends rather than means. According to Dewey, a worthwhile instructional sequence provided students with the opportunity to formulate and test hypotheses and thus engage in the reflective thinking process. Jacques Barzun (1991) states, "All that is new or seems new in Dewey . . . is the recommendation to make early instruction follow the pattern of scientific inquiry" (p. 56).

Dewey's ideas stimulated the progressive movement in American education. By the 1930s, the Progressive Education Association established the Committee on the Function of Science in General Education of the Commission on Secondary School Curriculum. The committees released their report as *Science in General Education* in 1938 (Commission on Secondary School Curriculum, 1937). Dewey's model of reflective thinking became a consistent theme and instruction model in that report. In a chapter on reflective thinking, the report had a section on "How the Science Teacher May Encourage Reflective Thinking." The section included the following recommendations:

- widening the range of problems that are real and stimulating to the student (p. 313)

- overcoming obstacles to the wide use of reflective thinking (p. 314)

- finding opportunity for students to work intensively on their problems (p. 315)

- working on problems encountered in meeting needs (p. 318)

- helping older students to become aware of the nature of reflecting thinking (p. 320)

- encouraging confidence in scientific methods (p. 324)

Underlying these general recommendations, one detects the model of scientific inquiry and provisions for students to engage actively in that process. The following list summarizes Dewey's instructional model. I synthesized this model from Dewey's statements and from *Science in General Education*.

Sensing perplexing situations	The teacher presents an experience where the students feel thwarted and sense a problem.
Clarifying the problem	The teacher helps the students identify and formulate the problem.
Formulating a tentative hypothesis	The teacher provides opportunities for students to form hypotheses and tries to establish a relationship between the perplexing situation and previous experiences.
Testing hypothesis	The teacher allows students to try various types of experiments, including imaginary, pencil-and-paper, and concrete experiments, to test the hypothesis.
Revising rigorous tests	The teacher suggests tests that result in acceptance or rejection of the hypothesis.
Acting on the solution	The teacher asks the students to devise a statement that communicates their conclusions and expresses possible actions.

Piaget's Model

Piaget's model of equilibration is the part of his overall developmental theory that relates to the process of learning. Equilibration is, in essence, the learning theory for Piaget's developmental psychology. Piaget's major statement on equilibration is *The Development of Thought* (1975). Piaget and others have discussed the equilibration process in their works (Nodine, Gallagher, and Humphreys, 1972; Piaget and Inhelder, 1969).

According to Piaget, intellectual development occurs through an adaptation in response to a discrepancy between the individual's current cognitive structure and a cognitive referent in the environment. Disequilibrium results from a discrepancy. Modification of intellectual structures brings the cognitive system back to equilibrium. The process of equilibration involves both maintenance and change of intellectual structures. That is, old structures continue while they are modified or combined with others to form new structures. Equilibration occurs at three major levels: (1) the total system, which Piaget described as a stage of development, a complete set of cognitive structures that are at equilibrium for a period of time; (2) a subsystem, such as the cognitive structures for classification, conservation, and numeration; (3) specific concepts, like color, objects, and organisms, which may be included in the aforementioned levels of organization.

In Piaget's model, organization and adaptation are the two processes that bring about equilibration. Organization is the maintenance of an internal order of the intellectual structure through the inherent tendency to systematize and integrate intellectual structures into coherent systems. This tendency results in the spontaneous transition to higher orders of intellectual complexity. These processes are further described in an educational context elsewhere (Bybee and Sund, 1982; Lawson and Wollman, 1975).

Adaptation is the process of changing the intellectual structure through interaction with the environment. This modification results in development of the cognitive structure. Intellectual adaptation consists of two processes that are simultaneous and complementary. These processes are accommodation and assimilation.

I can clarify the processes of accommodation and assimilation with an educational example. Assume that cognitively a student is at equilibrium with respect to particular events in the world, say, the motion of objects in response to the gravitational pull of the earth. That is, the student has a conception of the world that satisfactorily explains how and why objects roll downhill. The teacher confronts the student with a puzzling science demonstration, a discrepant event. For example, an object released on an inclined plane rolls up the plane rather than down. This instance of an object rolling uphill is associated with prior experiences of objects rolling downhill. The student tries to explain the event in terms of prior experiences (assimilation) and cannot do so. So the student has to change some cognitive schemes in order to explain the anomalous event. Accommodation occurs in the same time frame as the assimilation. The intellectual structure is somewhat readjusted to the external reality, the phenomenon of the demonstration. The student's prior conception of how objects respond to the earth's pull is changed, or a new explanation is developed that maintains the "gravity model" of objects rolling downhill. For instance, the student explains the event by another model (the object rolled uphill because it was powered by a rubber band).

I briefly summarized the process of equilibration for several reasons. The SCIS learning cycle, which I introduce next, is based on this model of learning. Much of the recent research on the cognitive sciences assumes a similar model. And the instructional model I propose assumes that students' construct knowledge in a similar manner.

The Atkin/Karplus Model

In the early 1960s, J. Myron Atkin and Robert Karplus (1962) proposed a systematic approach to instruction. Robert Karplus, Herbert Their, and their colleagues later used the instructional model, referred to as the learning cycle, in the SCIS. The learning cycle for SCIS was related to the psychological theories of Jean Piaget. The classic article "Messing About in Science" by David Hawkins (1965) described a teaching model using symbols of circle, triangle, and square. In general, the symbols represented phases of an instructional model that included unstructured exploration and multiple programmed experiences and didactic instruction.

In past decades, the SCIS learning cycle has undergone elaboration, modification, and application to different educational settings. In addition, recent analyses of elementary programs indicate that SCIS was one of the more effective programs (Shymansky, Kyle, and Alport, 1983). These positive effects on learning relate at least in part to the learning cycle. Most recently, the SCIS learning cycle has been used as central to a theory of instruction (Lawson, Abraham, and Renner, 1989) (see Figure 8.1).

Using the learning cycle creates situations in which the processes of assimilation, accommodation, and organization occur (Renner and Lawson, 1973; Lawson, Abraham, and Renner, 1989). The three phases and the normal sequence of the SCIS learning cycle are exploration, invention, and discovery. Exploration refers to the relatively unstructured experiences in which students gather new information. This phase predominantly involves the process of assimilation. Invention refers to a formal statement, often the definition, of a new concept. Following the exploration, the invention phase begins the process of accommodation that allows interpretation of newly acquired information through the restructuring of prior concepts. The discovery phase involves application of the new concept to another, novel situation. During this phase, the learner continues to move closer to a state of equilibrium and to a new level of cognitive organization (integration of the new concept with related concepts).

A number of studies have shown that the SCIS learning cycle has many advantages when compared with other approaches to instruction, specifically the transmission model of teaching. These studies are summarized in Abraham and Renner (1986). Jack Renner and his colleagues (Renner, Abraham, and Birnie, 1985; Abraham and Renner, 1986; Renner, Abraham, and Birnie, 1988) have investigated, respectively, the form of acquisition of information in the learning

FIGURE 8.1 *The Atkin/Karplus Instructional Model*

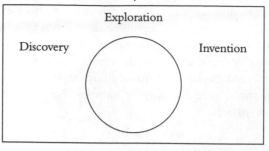

cycle, the sequencing of phases in the learning cycle, and the necessity of each phase of the learning cycle. These studies have generally supported use of the SCIS learning cycle as originally designed by Atkin and Karplus. Research on discovery, guided discovery, and statement-of-rule learning (Egan and Greeno, 1973; Gagne and Brown, 1961; Roughead and Scandura, 1968) supports the "sequencing and necessity" conclusions drawn by Renner and his colleagues.

Although the form and structure of the Atkin/Karplus learning cycle have undergone little revision, researchers have offered different interpretations of each phase of the cycle. Abraham and Renner (1986) refer to the exploration phase as "gathering data." This interpretation provides more structure for the assimilation of new material and restricts Atkin and Karplus' original notion that exploration should provide students with common experiences, regardless of whether those experiences involve gathering data in a laboratory sense.

Lawson (1988) describes the invention phase as "concept introduction," suggesting that because the new concept is not fully developed during this phase, the learner does not truly invent the concept. Lawson further suggests that the appropriate label for this phase may be "term introduction," since only the vocabulary associated with the new concept is learned at this point.

Renner renamed the discovery phase "expansion," taken from the idea that the learner actually expands on the new concept and is involved in the Piagetian process of organization. Lawson (1988) suggested a more restrictive interpretation of this phase when he used the words "concept application." One should be aware, however, that in applying the concept to new situations, the learner may still be in the process of restructuring or reconstructing the concept.

Design Requirements for an Instructional Model

Design for an instructional model must bridge the everyday and scientific thinking of students and teachers (Hawkins and Pea, 1987). The model has to have a practical quality for teachers and provide understandable features for students. The following statements describe some of the essential design requirements for an instructional model:

1. The model should incorporate the essential features of the three phases of the SCIS learning cycle in the traditional sequence. The research base for this model is substantial and should not be ignored.

2. The model must have an initial stage that engages the learner and brings about disequilibrium in the student. This stage should be designed to gain the student's attention on critical aspects of an experience and to focus thinking on concepts deemed important.

3. The model must help the learner integrate new knowledge with prior knowledge. In order to understand any new concept, the learner must, in a meaningful way, connect that concept to other concepts in science and technology as well as to new and old experiences.

4. The model must allow for student–student as well as student–teacher interaction. Social interaction among students encourages argument and opportunities to formulate new explanations for experiences.

5. The model must help the learner through the process of conceptual change in a "generic" fashion that includes, but is not tied to, specific misconceptions. That is, the instructional model must be applicable to many science concepts. Designing an instructional model unique to each science concept seems complex, inefficient, and fails to recognize the fact that science teachers must interact with twenty-five or thirty students at a time. This requirement is both crucial and challenging. The research on conceptual change necessarily focuses on specific concepts, but teachers cannot anticipate and recognize all of the student conceptions associated with a particular topic or curriculum. The model must therefore allow for explanations generated by students as well as ample time for discussion of those explanations.

6. Although allowing for conceptual change, the model must be manageable for the teacher. It is generally not feasible for a teacher, prior to any lesson, to ascertain each student's prior conceptions and then structure instruction accordingly. In the future, technology may help overcome this problem. But for now, teachers will confront classrooms with considerable variation in conceptual understanding. Thus, the model should include strategies that are designed to maximize the benefit for individual students and also that are usable for the teacher in the typical classroom environment.

7. The model must be intelligible and apparent to both teachers and students. Teachers who initially use an instructional model are themselves engaged in a process of conceptual change. Most science teachers have models of teaching and learning that probably vary with the one presented here. The model should therefore be easily recognizable. Students also should know what phase of the model they are in and what is expected of them during that phase so that they can engage in appropriate behaviors.

8. The model must be applicable via a variety of teaching strategies, including traditional laboratories and new educational technologies.

9. The model must incorporate the traditional processes of scientific inquiry and technological design.

10. The phases of the model must be identifiable by words easily understood and applied by practicing science teachers.

An Instructional Model for Contemporary Science Education

The instructional model proposed here fulfills the criteria described in the previous section. The model has five phases and includes structural elements in common with the original Atkin/Karplus learning cycle. The five phases are engagement, exploration, explanation, elaboration, and evaluation. The middle three phases are fundamentally equivalent to the three phases of the Atkin/Karplus learning cycle. In reading the description of the model proposed here, you should note the variations from the original model and the different appropriate activities and procedures for each phase. These interpretations serve to define the model every bit as much as the skeletal structure of the model.

Curriculum developers and classroom teachers can apply the model at several levels. The model can be the organizational pattern for a year-long program, for units within the curriculum, and for sequences of daily lessons. I recommend, however, that initially developers and teachers use the model clearly and consistently at only one of these levels. Although curriculum developers who are familiar with the model may be able to utilize nested cycles (cycles within cycles) or interwoven cycles (exploring one concept while expanding on another), apprising teachers and students of that nesting or interweaving would neglect the "intelligible and apparent" criterion.

The instructional model is based on a constructivist view of learning. Constructivism is a dynamic and interactive model of how humans learn. Using this approach, students redefine, reorganize, elaborate, and change their initial concepts through self-reflection and interaction with their peers and their environment. Learners interpret objects and phenomena, and internalize those interpretations in terms of their current conceptional understanding. The objective in a constructivist program is to challenge students' current conceptions by providing discrepant events, data that conflict with students' current thinking, or experiences that provide an alternative way of thinking about objects and phenomena. When an activity challenges students' conceptions, there must be an opportunity for students to reconstruct a conception that is more adequate than their original conception. This takes time and a planned sequence of instruction. The instructional model outlined in the next sections provides the time, opportunity, and structure necessary for such learning to occur. Table 8.2 summarizes the instructional model.

Efforts to translate constructivist research into classroom practice is evident in the use of the 5 Es instructional model. Instructional procedures have been developed to help students through the process of conceptual change (Bruer, 1993; Driver, 1989; Driver et al., 1994; Driver, Guesne, and Tiberghien, 1985; Hewson, 1984; McGilly, 1994; Kyle, Abell, and Shymansky, 1989; Osborne and Freyberg, 1985). Many of these efforts have improved student understanding in the physical sciences (White and Gunstone, 1992; Hewson, 1984; Hewson, 1986; Joshua and Dupin, 1987; Minstrell, 1984; Minstrell, 1989; White, 1985) and the biological sciences (Anderson and Roth, 1987). A pedagogical tool known as concept mapping (Novak and Gowin, 1984) has been used to allow students to articulate their current conceptions. The use of concept mapping has been shown to improve achievement in a variety of subjects (Novak, 1987; Tisher, 1985). Berkheimer and Anderson (1989) and Driver and Oldham (1986) have outlined constructivist approaches to curriculum development. These approaches rely heavily on determining students' prior knowledge and structuring instruction accordingly.

The following sections present brief descriptions of the five phases of the instructional model. Curriculum developers and classroom teachers can apply the instructional model in different disciplines and with varied teaching strategies.

Engagement

The first phase is to engage the student in the learning task. The student mentally focuses on an object, problem, situation, or event. The activities of this phase should make connections to past and future activities. The connections depend on the learning task and may be conceptual, procedural, or behavioral.

Asking a question, defining a problem, showing a discrepant event, and acting out a problematic situation are all ways to engage the students and focus them on the instructional activities. The role of the teacher is to present a situation and identify the instructional task. The teacher also sets the rules and procedures for the activity. The experience need not be long or complex; in fact, it should probably be short and simple.

Successful engagement results in students being puzzled by, and actively motivated in, the learning activity. Here the word *activity* refers to both a constructivist and a behavioral approach, that is, the students are mentally and physically active.

Exploration

Once the activities have engaged students, they need time to explore their ideas. Exploration activities are designed so that all students have common, concrete experiences upon which they continue building concepts, processes, and skills. If engagement brings about disequilibrium, exploration initiates the process of equilibration. This phase should be concrete and meaningful for the students.

The aim of exploration activities is to establish experiences that teachers and students can use later to formally introduce and discuss scientific and technological concepts, processes, or skills. During the activity, the students have time in which they can explore objects, events, or situations. As a result of their mental and

TABLE 8.2 *The 5 Es Instructional Model: Examples of Student Behaviors and Teacher Strategies*

Student Behavior	*Stage of the Instructional Model*	*Teaching Strategy*
	Engage	
Asks questions such as, Why did this happen? What do I already know about this? What can I find out about this? How can this problem be solved?	Initiates the learning task. The activity should make connections between past and present learning experiences, and anticipate activities and organize students' thinking toward the learning outcomes of current activities.	Creates interest.
		Generates curiosity.
		Raises questions and problems.
Shows interest in the topic.		Elicits responses that uncover students' current knowledge about the concept/topic.
	Explore	
Thinks creatively within the limits of the activity.	Provide students with a common base of experiences within which current concepts, processes, and skills are identified and developed.	Encourages students to work together without direct instruction from the teacher.
Tests predictions and hypotheses.		Observes and listens to students as they interact.
Forms new predictions and hypotheses.		
Tries alternatives to solve a problem and discusses them with others.		Asks probing questions to redirect students' investigations when necessary.
Records observations and ideas.		Provides time for students to puzzle through problems.
Suspends judgment.		
Tests ideas.		Acts as a consultant for students.
	Explain	
Explains possible solutions or answers to other students.	Focus students' attention on a particular aspect of their engagement and exploration experiences, and provide opportunities to demonstrate their conceptual understanding, process skills, or behaviors. This phase also provides opportunities for teachers to introduce a concept, process, or skill.	Encourages students to explain concepts and definitions in their own words.
Listens critically to other students' explanations.		Asks for justification (evidence) and clarification from students.
Questions other students' explanations.		
Listens to and tries to comprehend explanations offered by the teacher.		Formally provides definitions, explanations, and new vocabulary.
Refers to previous activities.		Uses students' previous experiences as the basis for explaining concepts.
Uses recorded observations in explanations.		

TABLE 8.2 *The 5 Es Instructional Model: Examples of Student Behaviors and Teacher Strategies (continued)*

Student Behavior	Stage of the Instructional Model	Teaching Strategy
	Elaborate	
Applies new labels, definitions, explanations, and skills in new, but similar, situations.	Challenge and extend students' conceptual understanding and skills. Through new experiences, the students develop deeper and broader understanding, more information, and adequate skills.	Expects students to use vocabulary, definitions, and explanations provided previously in new context.
Uses previous information to ask questions, propose solutions, make decisions, design experiments.		Encourages students to apply the concepts and skills in new situations.
Draws reasonable conclusions from evidence.		Reminds students of alternative explanations.
Records observations and explanations.		Refers students to alternative explanations.
	Evaluate	
Checks for understanding among peers.	Encourage students to assess their understanding and abilities and provide opportunities for teachers to evaluate student progress.	Refers students to existing data and evidence and asks, What do you already know? Why do you think . . .?
Answers open-ended questions by using observations, evidence, and previously accepted explanations.		Observes students as they apply new concepts and skills.
Demonstrates an understanding or knowledge of the concept or skill.		Assesses students' knowledge and/or skills.
Evaluates his or her own progress and knowledge.		Looks for evidence that students have changed their thinking.
Asks related questions that would encourage future investigations.		Allows students to assess their learning and group process skills.
		Asks open-ended questions such as, Why do you think . . .? What evidence do you have? What do you know about the problem? How would you answer the question?

physical involvement in the activity, the students establish relationships, observe patterns, identify variables, and question events.

The teacher's role in the exploration phase is that of facilitator or coach. The teacher initiates the activity and allows the students time and opportunity to investigate objects, materials, and situations based on each student's own ideas of the phenomena. If called upon, the teacher may coach or guide students as they begin constructing new explanations. Use of tangible materials and concrete experiences are essential in the exploration phase.

A portion of the exploration phase should center on cooperative learning (Johnson and Johnson, 1987; Johnson, Johnson, and Holubec, 1986; Johnson, Johnson, and Maruyama, 1983). The opportunity for students to interact, discuss, and even argue in a constructive environment and about goal-centered activities enhances the possibility that their current concepts will be challenged and other ideas will be evident as they reconstruct their ideas (Kuhn, 1992; Vygotsky, 1962, 1978; Smith, Carey, and Wiser, 1985; Champagne, 1987).

Explanation

Explanation means the act or process in which concepts, processes, or skills become plain, comprehensible, and clear. The process of explanation provides the students and teacher with a common use of terms relative to the learning experience. In this phase, the teacher directs student attention to specific aspects of the engagement and exploration experiences. First, the teacher asks the students to give their explanations. Second, the teacher introduces scientific or technological explanations in a direct and formal manner. Explanations are ways of ordering and giving a common language for the exploratory experiences. The teacher should base the initial part of this phase on the students' explanations and clearly connect the explanations to experiences in the engagement and exploration phases of the instructional model. The key to this phase is to present concepts, processes, or skills briefly, simply, clearly, and directly, and then continue on to the next phase.

The explanation phase is teacher-directed. Teachers have a variety of techniques and strategies at their disposal. Educators commonly use verbal explanations, but there are numerous other strategies, such as video, films, and educational courseware. This phase continues the process of mental ordering and provides words for explanations. In the end, students should be able to explain their experiences to each other and to the teacher.

Elaboration

Once the students have an explanation of their learning tasks, it is important to involve them in further experiences that apply, extend, or elaborate the concepts, processes, or skills. Some students may still have misconceptions, or they may only understand a concept in terms of the exploratory experience. Elaboration activities provide further time and experiences that contribute to learning.

Audrey Champagne (1987) discusses an example of the elaboration phase:

Students engage in discussions and information-seeking activities. The group's goal is to identify and execute a small number of promising approaches to the task. During the group discussion, students present and defend their approaches to the instructional task. This discussion results in better definition of the task as well as the identification and gathering of information that is necessary for successful completion of the task. The teaching model is not closed to information from the outside. Students get information from each other, the teacher, printed materials, experts, electronic databases, and experiments which they conduct. As a result of participation in the group's discussion, individual students are able to elaborate upon the conception of the tasks, information bases, and possible strategies for its [the task's] completion. (p. 82)

Note the use of interactions within student groups as a part of the elaboration process. Group discussions and cooperative learning situations provide opportunities for students to express their understanding of the subject and receive feedback from others who are very close to their own level of understanding.

The elaboration phase also is an opportunity to involve students in new situations and problems that require the application of identical or similar explanations. Generalization of concepts, processes, and skills is the primary goal of the elaboration phase.

Evaluation

At some point, it is important that students receive feedback on the adequacy of their explanations. Informal evaluation can occur from the beginning of the teaching sequence. The teacher can complete a formal evaluation after the elaboration phase. As a practical educational matter, science teachers must assess educational outcomes. This is the phase in which teachers administer tests to determine each student's level of understanding. This also is the important opportunity for students to use the skills they have acquired and evaluate their understanding. The instructional model proposed here is closely aligned with the actual processes involved in the scientific and technological enterprise. In science and technology, the methods of scientific inquiry and strategies of design are excellent means for students to test their explanations and solutions. This is, after all, congruent with science and technology. How well do the students' explanations and solutions stand up to review by peers and teachers? Is there a need to reform ideas based on experience? Table 8.3 provides another summary of the instructional model and suggests parallels to scientific inquiry and technological problem solving.

Some Reflections on the Instructional Model

In my work at BSCS, I have used this instructional model in curriculum materials for the elementary school—*Science for Life and Living: Integrating Science, Technology,*

TABLE 8.3 *The Instructional Model and Contexts for Science and Technology*

A Scientific Context for the Instructional Model	An Instructional Model	A Technological Context for the Instructional Model
The student has questions about the natural world.	*Engagement:* This phase of the instructional model initiates the learning task. The activity should make connections between past and present learning experiences and anticipate activities and focus students' thinking on the learning outcomes of current activities. The student should become mentally engaged in the concept, process, or skill to be explored.	The student has a problem about a human adaptation to the environment.
The student uses scientific inquiry to answer the questions. Scientific approaches such as stating an appropriate question, making observations, doing an investigation, gathering and analyzing data are all part of this phase in science teaching.	*Exploration:* This phase of the teaching model provides students with a common base of experiences within which they identify and develop current concepts, processes, and skills. During this phase, students may use cooperative learning to explore their environment or manipulate materials.	The student uses different strategies to solve the problem. Engineering approaches, such as recognizing constraints and criteria, analyzing costs, risks, benefits, and designing prototypes are part of this phase of teaching.
The student proposes answers to the questions.	*Explanation:* This phase of the instructional model focuses students' attention on a particular aspect of their engagement and exploration experiences, and provides opportunities for them to verbalize their conceptual understanding or demonstrate their skills or behaviors. This phase also provides opportunities for teachers to introduce a formal label or definition for a concept, process, skill, or behavior.	The student proposes a solution to the problem.

TABLE 8.3 *The Instructional Model and Contexts for Science and Technology (continued)*

A Scientific Context for the Instructional Model	*An Instructional Model*	*A Technological Context for the Instructional Model*
The student applies the proposed answers to new situations in an effort to generalize the explanation.	*Elaboration:* This phase of the teaching model challenges and extends students' conceptual understanding and allows further opportunity for students to practice desired skills and behaviors. Cooperative learning is appropriate used in this stage. Through new experiences, the students develop deeper and broader understanding, more information, and adequate skills.	The student tests the solution in different contexts.
The student and teacher determine the adequacy of the explanation.	*Evaluation:* This phase of the teaching model encourages students to assess their understanding and abilities, and provides opportunities for teachers to evaluate student progress toward achieving the educational objectives.	The student evaluates the solution in terms of the criteria and constraints.

and Health; for the middle school—*Middle School Science and Technology;* and for the high school—*Biological Science: A Human Approach.*

Let me describe several critical issues about instructional models. The first issue has to do with the meaning an activity within the instructional sequence has for students. Students derive personal meaning from three types of experiences. The general terms *physical, psychological,* and *social* describe the types of meaning an experience has for students. An activity can have meaning for students because objects and events are physically close. Objects and events have meaning for individuals simply because they are close and involved. Placing an unknown object in a student's hand increases the meaning of that object for the student. Having hands-on experiences and engaging in problem-solving activities both have a dimension of personal meaning. There is a second dimension of psychological meaning. Some objects and events are interesting and engaging for students. Dinosaurs, plants, and the solar system are all examples of psychologically interesting things for children. Instruction can use the initial interest in these areas to develop concepts, such as time, cycles, and scale. Finally, there is a social aspect of meaning. This is the dimension that most individuals associate with meaning. Educators often equate meaning with relevance or the timeliness of issues. In some cases, science-related social issues, such as population growth, environmental pollution, or resource depletion are meaningful to students. However, assuming these are meaningful to students just because they are timely, and even critically important, is not always a correct assumption from a learning point of view. Obviously, combining all three dimensions of meaning in educational experiences certainly enhances the possibilities of learning.

There is a second critical issue regarding instructional models. An instructional model is probably necessary but not sufficient for the process of conceptual change. That is, the careful structuring and sequencing of activities helps tremendously to bring about conceptual change. There remains, however, a critical interaction among students and between teachers and students that completes the process in the context of classrooms and schools. A carefully structured sequence of activities enhances the possibilities of learning, but it does not ensure learning. The careful probing by teachers, subtly challenging the students, and knowing when to provide a hint or clue that will help the student reconstruct an idea are all interpersonal dimensions of instruction that cannot be adequately accommodated by a set of activities. In short, the burden of learning is too heavy to place solely on an instructional model. The teacher is essential to complete the process of conceptual change.

Finally, the instructional model provides an educational bridge from the students' current conceptions or misconceptions to current scientific concepts. You may note the careful structuring and sequencing that allows students to identify explanations and their adequacy. Although honoring students' ideas, it is also the responsibility of science curriculum and instruction to introduce and help students learn scientific concepts and processes. The design of the 5 Es model is such that this occurs at the midpoint of the instructional sequence. The elaboration and evaluation phases provide time and opportunity for students and teachers to assess their own understanding against those of science.

Conclusion

This chapter describes an instructional model designed for several BSCS programs. An instructional model brings coherence to different teaching strategies, provides connections among educational activities, helps science teachers make decisions about interactions with students, and contributes to students' development of scientific and technological literacy.

The phases for the instructional model date back to the ideas of Johann Friedrich Herbart, John Dewey, Jean Piaget, and J. Myron Atkin and Robert Karplus. Although the model dates to the ideas of these individuals, it has been modified and updated based on contemporary research and practical issues of science teaching. The instructional model has constructivism as a theoretical foundation but recognizes the critical role of classroom teachers, who must make myriad decisions about their students.

The instructional model consists of five phases designed to facilitate the process of conceptual change. The actual application of the phases in curriculum and teaching may not be as clear and easy as outlined here; still, the model should contribute to better, more consistent, and coherent instruction. The instructional sequences include

- *Engagement*. In the first phase, the teacher designs experiences intended to make connections with current concepts and skills and to bring into question the adequacy of those concepts and skills.

- *Exploration*. In this phase, the teacher uses activities and social interaction (cooperative learning) to help students begin constructing more adequate concepts and developing better skills.

- *Explanation*. In this phase, the students have an opportunity to articulate their ideas, and the teacher helps students clarify their ideas through scientific and technological terms and concepts.

- *Elaboration*. In this phase, the teacher provides activities based on the same concepts and skills, but there is a new and different context. The students must expand or generalize their new conceptions to the different experiences.

- *Evaluation*. In this phase, the teacher uses a variety of assessments to determine the students' conceptual understanding and level of skill development. This phase also is an opportunity for students to test their understanding and skills.

Four general factors support this instructional model: (1) educational research on conceptual change, (2) congruence of the model with the general processes of scientific inquiry and technological design, (3) utility of the model for designing and developing curriculum materials, and (4) practical use by science teachers.

Reforming science education and transforming purposes to practices will only occur when science teachers improve instructional practices. Implementing systematic and coherent approaches to science teaching such as the 5 Es model represents a tremendous advance in the reform of science education.

9

Implementing a Standards-Based Systemic Reform

The immense journey of contemporary educational reform began with declarations that our nation was at risk, proceeded to the major educational aims of *Goals 2000,* and progressed to the development of standards. Whether local, state, or national, the development of standards for science education represents an essential step; and contrary to some views, the step to develop standards was one of the easiest, not the hardest. The next steps on the journey involve improving instructional materials, changing teaching practices, and enhancing assessment strategies. Central to these steps is the professional development of science teachers, because the ultimate reform of science education must occur in classrooms and in the interactions among science teachers and students. The next steps toward achieving scientific literacy will cost more, take more time, involve more people, and require coordination and coherence within the educational system. The payoff for these increases of costs, time, personal involvement, and systemic reform, however, will be substantial in terms of moving learners to higher levels of scientific literacy.

This chapter first examines some ideas and assumptions about standards, introduces some ideas related to large-scale reform, then elaborates the idea of systemic reform in science education. Finally, I provide insights from the area of large-scale policy.

Understanding What Science Education Standards Are

Before discussing the meaning of standards, it will be helpful to review several of the *National Science Education Standards.* Following are listed two standards from content (inquiry and life science) and one each from teaching and assessment. I include these four standards because one must view all content standards and not examine only subject matter standards (physical, life, earth and space science, inquiry, technology, science in personal and social perspectives, history and nature of science); the standards must be viewed as an integrated whole, which means that standards on teaching, professional development, assessment, program, and system should be included with content; and each of the four standards has a different form and function and will provide different meanings for the following discussions.

Standard for Science as Inquiry

As a result of their activities in grades five through eight, all students should develop

- the abilities of scientific inquiry
- understandings about scientific inquiry

Standard for Life Science

All students should also develop an understanding of

- structure and function in living systems
- reproduction and heredity
- regulation and behavior
- populations and ecosystems
- diversity and adaptations of organisms

Standard for Science Technology

Teachers of science should plan an inquiry-based science program for their students that incorporates the following elements:

- a framework of year-long and short-term goals for students
- science content and curricular design that meets the particular interests, knowledge, skills, and experiences of students
- teaching strategies that support the development of student understanding and nurture a community of science learners
- collaborative work within and across disciplines and grade levels

Standard for Assessment

Student achievement and opportunity to learn science must both be assessed.

- Achievement data collected focuses on the science content that is most important for students to learn.
- Opportunity-to-learn data collected focuses on the most powerful indicators of students' opportunity to learn.
- Equal attention must be given to the assessment of opportunity to learn and the assessment of student achievement.

This discussion of standards applies to standards at state and local levels as well as national standards.

The use of standards in educational reform is at one time an excellent, much needed idea and a complex and ambiguous innovation. Standards have different meanings and uses in the educational system. For example, they may represent commitment and vision, and they may be used as exemplars, rules, and uniform measures. Gary Sykes and Peter Plastrik (1993) have written an excellent publication on *Standard Setting as Educational Reform*. Many of their ideas are represented in this section. Sykes and Plastrik identify three models of reform: the systemic reform, the professional model, and the reform network model. My discussion emphasizes the systemic reform model, which focuses on student learning and the use of standards to define curriculum content and assessment. The systemic reform model builds on the original ideas of Smith and O'Day (1991), Cohen and Spillane (1993), and the first standards documents developed by the NCTM (1989). The connection between standards and systemic reform was clearly established with the NSF program solicitation for statewide systemic initiatives in science, mathematics, and technology education. That solicitation was issued in 1990 under the leadership of Luther Williams, Associate Director for the Division on Education and Human Resources.

Standards as a Call to Action

The term *standards* has several meanings, all of which have appropriate uses and contexts in science education. Individuals commonly associate the idea of standards with a flag, banner, or ensign that serves as a point or symbol to rally military action. In this view, standards represent commitment to a territory, advancement toward a goal, and aggressive action to achieve an objective. The movement to set standards for science education certainly fits this use of the term. Setting standards, such as those listed at the beginning of this section, signals commitment, action, and the need to achieve particular objectives. In the vast territory of content that might be included in the science curriculum and used to define achievement, the first two standards listed (inquiry and life science) are examples of an agreed-upon commitment by the science education community. To be more concrete, all science content does not have equal weight in decisions about what is included in the curriculum, what is taught, and what is assessed. For those concerned with science education—curriculum developers, policy makers, funding agencies, commercial publishers, and school personnel—the standards should have more weight when it comes to decisions relating to school science curriculum and assessment.

Standards as a Measure of Value

Standards also means an acknowledged measure of comparison for quantitative and qualitative value. One aspect of this is the quantitative, setting a unit of measure, such as Greenwich Mean Time or the length of a meter that serves as a point of reference for other measures of, for example, distance. Here, measurement becomes an important aspect of standards, and thus there is a close connection between assessment (see the fourth standard in the preceding list) and the process of setting standards. Many educators have not yet realized that there is a close connection between content standards and performance standards. It is one thing to say that all students completing grade 8 should develop understandings of structure and function in living systems and yet another to provide performance standards and measurements for what counts as achieving that content standard. This view of standards as measurement underscores the importance of assessment and the power of standards as expectations of achievement.

Most educators acknowledge that the quantitative measurement of standards presents significant but not insurmountable problems. Certainly, groups such as Educational Testing Service, the College Board, and various state and local assessment agencies will endeavor to design instruments quantifying levels of achievement for national, state, and local standards. The standards also have value in a more qualitative dimension of measurement: they act as authoritative exemplars of content and opportunities to learn science. Whether at the national, state, or local level, standards act as statements for discussion, argument, and disputation. Such a view of standards recognizes that they are not doctrine but rather statements that must be interpreted and analyzed for them to make sense and have meaning. Such a view of continual and widespread interpretation reminds one of biblical exegesis and Talmudic commentaries, and presents a stark contrast to the view of clearly defined nondisputable requirements for content, teaching, and assessment. Personally, I find this idea and use of standards to have rich implications for the professional development of all science educators. Imagine meetings and seminars discussing the subtleties of *Standards* like those in the preceding list as professionals search for meaning and eventual implementation of the standards. This view presents a poignant contrast to "Just give me the activities," "Tell me what to do," "Will the tests include standards?" and "*They* have to change."

Standards as an Assessment

Another view of standards suggests a more abstract level of excellence, attainment, and quality ascribed to the entire set of standards. In an educational context, we talk about higher levels of achievement for all students or being first in the world in science and mathematics by the year 2000. Standards will define what we mean by higher levels of achievement and could clarify our place on international assessments. I must say that the latter expresses political aspirations more than educational realities.

Standards as Reinforcing and Balancing Feedback

Using standards to assess changes in curriculum, instruction, assessment, programs, and the educational system represents their use as feedback. Feedback provides two basic dynamics: reinforcing (positive feedback) and balancing (negative feedback). Reinforcing feedback processes results in growth and amplifies situations, sometimes in detrimental ways. Escalation in an arms race and accelerating declines in sales of a product represent situations where reinforcing feedback is probably at work. Relative to standards, one significant fear is that they will not be interpreted and appropriately implemented, thus reinforcing the status quo and contributing to further decline in science education. This situation can result from reviewing only a single aspect of the standards, for example, subject matter in one discipline or one grade level. From this perspective, science educators could conclude that change is not necessary, or overinterpret some aspects of the standards and suggest they mean more than they actually do, or underinterpret or ignore other aspects of the standards. The net result of over- or underinterpretation is a deviation from the vision set forth in the standards. This is a potential problem of positive feedback.

Balancing feedback operates when there is a goal such as maintaining a specific temperature in a room or speed in an automobile. In science education, the goal can be explicit, such as aligning curriculum, instruction, and assessment with standards. In the case of standards-based reform, the standards can act to balance processes of change and self-correct when changes do not align with goals. As we progress toward various goals, such as improving assessments, standards provide feedback and correction. Balancing feedback can cause difficulties because it is often hard to see, so there is little evidence for change. For example, the temperature in the room stays about the same. What standards will do is metaphorically reset the thermostat and initiate new parameters and processes for balancing feedback.

To the question, What are standards for science education? I have to answer that standards are several things, including commitments to certain goals for science education, measurements of attainment, context for professional discourse, and criteria to assess curriculum, teaching, and assessments. Standards have different meanings, orientations, and uses depending on the context.

With some agreement on the primary structures for standards and their uses to rally, guide, and measure, the actual translation of standards within the science education system, including everything from national assessments to classroom practices, is a complex process involving issues such as allegiance of diverse groups, political commitments, and financial support. Underlying this process of translating standards to programs and practices, one finds enduring, and I think essential, tensions between the authority of society and the autonomy of individuals (Sykes and Plastrik, 1993). In other words, does a society have the right to set and enforce standards for science education that challenge the autonomy of individuals and groups, such as teacher educators, curriculum developers, assessment specialists, and classroom teachers? Standards for science education do represent a societal consensus on what students should know and be able to do. And the science education community should not forget that the contemporary reform emerged from a perceived need to change.

Standards have the potential to improve science education and move us toward higher levels of scientific literacy. Standards for science education can

- serve as purposes toward which all individuals in the community should strive

- represent exemplary content, teaching, and assessments

- provide principles that guide development of programs and influence practices

- coordinate and align various components of the science education system

- create a shared set of criteria for improving science education and achieving scientific literacy

- provide feedback in national, state, and local progress toward the goal of scientific literacy

Several Assumptions Underlying the Use of Science Education Standards in Reform

Although national standards receive considerable attention, since about 1990 states and local school districts also have used standards as primary policies to direct the course of reform for education, including science education. In 1989, when the NCTM first used the term *standards* in the context of educational reform, it was innovative and somewhat controversial. By 1995 educators commonly used the term with considerable ambiguity. The unfortunate result of this ambiguity is a failure to recognize the multiple meanings and complexities and to fall prey to prior, often naive, conceptions of the idea. Especially problematic is a widespread failure to recognize the assumptions associated with standards-based reform. Paul LeMahieu and Helen Foss (1994) are particularly insightful concerning the assumptions of standards-based reform and the implications of those assumptions for policy and practice. Following are the seven assumptions from the LeMahieu and Foss article. In the next section, I discuss several points related to their essay.

Assumption 1. Standards-based reform requires that considerations of what students should know, be able to do, and be disposed to do should be placed at the center of reform efforts.

Assumption 2. Standards-based reform makes public the goals, expectations, and aspirations for individuals and the system and provides a focus for their efforts that will improve the effectiveness of the system.

Assumption 3. The articulation and widespread dissemination of standards will provide adequate guidance to ensure the transformation of teaching and learning.

Assumption 4. Standards-based reform will recouple effort and accomplishment.

Assumption 5. Standards-based reform permits an appropriate balance in who defines the educational experiences of students.

Assumption 6. Standards-based reform requires and promotes a necessary climate of trust among various elements of the educational system.

Assumption 7. Standards-based reform will enable a more appropriate and equitable allocation of funding and resources according to desired outcomes and needs.

A Focus on What Students Know and Are Able to Do

Listen to any group discussing the issue of reform, and one hears a list of familiar concerns including governance, administrative support, school climate, student attitudes, and testing. All such issues constitute important aspects of reform, but there is often a distinct lack of focus on the central issues, namely, teachers and teaching, and students and learning. Standards for science education send a clear signal that first and foremost all those in science education should place learners at the center of reform efforts. As LeMahieu and Foss suggest, "To fail to do so is to dictate that the winds of change will blow powerfully through the halls of our schools, but unfortunately not into any of the classrooms."

A Change from Inputs to Outcomes

By developing and implementing standards, we have changed the emphasis of educational reform from inputs to outcomes. The term *inputs* refers to such factors as length of school day and year, number of science courses required for graduation, time-on-task, adapting new curriculum, textbooks, use of computers, and so on. In implementing these inputs of more time and better materials, school personnel assumed there would be increased student learning. But we did not focus as clearly on outcomes, namely, what did we expect students to know and be able to do? Now standards-based reform shifts our attention to outcomes and requires an alignment of inputs to achieve the outcomes described in standards. Clearly, we need excellent inputs and clear expectations of outcomes, and both of these complementary aspects of education are evident in this reform. One result of the emphasis only on inputs was varying levels of achievement among states, schools, and students. This variation was acute particularly among traditionally underrepresented groups, and subsequently the approach worked against one other goal of American education, namely, equity. Now, through a process of reviews and consensus, the science education community established outcomes—standards for what we expect all learners to know and be able to do. We have fixed the outcomes and expect *all* students to achieve the same high standards. Inputs such as curriculum and instruction will have to vary according to student needs and the content defined in standards. Educators are just beginning to realize the significance and the implications of the change in educational orientation from inputs to outcomes. This change and the implications for instructional materials represents one of the most significant changes for curriculum reform in the history of American education (Schubert, 1993).

A Change from Single, Isolated Solutions to Systemic Reform

Presentations on the *Standards* (National Research Council, 1996a) bring predictable questions. Science teachers ask, Are there curriculum materials aligned with standards? Will tests change? Will my administration support the proposed change? Administrators ask, What do we have to do? How long will it take to implement standards-based programs? How much will it cost? Parents ask, Will my child learn the basics? Such questions are predictable because they emerge from concerns about how the standards will directly affect those who ask the questions. The questions also affirm the systemic perspective that has developed in parallel with the formulation of standards for science education.

Intuitively, most of us recognize the systemic nature of education. The foregoing questions reveal an understanding that in a complex system, you cannot just change one thing because everything is connected to everything else. To approach this reform in a systemic way, we will, for example, have to begin thinking and acting as though education is a system with varied interactive components described by uncertain boundaries and with feedback coordinating and regulating the activities of different components. The profound changes implied by instituting standards-based education can be enhanced by the systemic approaches, but we will all have to begin perceiving and behaving in terms of science education as a system (Senge, 1990a, 1990b).

A Public Review of Standards

As a note of contrast, consider the curriculum reform in science education that occurred in the 1950s and 1960s. Although the public had some awareness of the various curriculum projects, it is safe to say there was not the extensive review and input by various publics that we have witnessed in the contemporary development and implementation of standards for science education. The public nature of standards setting assumes the important inclusion of systemic components and tacitly acknowledges that many aspects of the system are accountable for achieving the aspirations defined in standards.

A Translation of Standards to Curriculum, Instruction, and Assessment

The process of establishing standards assumes there will be changes in school science programs and practices to help students achieve scientific literacy as defined by the standards. Needless to say, this assumption has already been challenged. First, it seems the dominant perceptions in the interpretation of standards are grounded in current programs and practices. Second, there is a clear reluctance to study and understand the standards as a total, integrated set of policies. Third, there is the ever-present view, based on a reluctance to change, that "we are already doing that." The latter is, of course, reinforced by publishers claiming that their products "meet the standards." Finally, one must acknowledge the comprehensive changes implied by standards for science education.

A Shared Responsibility for Decisions About Programs and Practices

Using the contemporary terms of centralized and decentralized approaches, the standards present a more centralized approach to the identification and clarification of outcomes and aspirations for students, that is, defining what they should know and do. In the past, such outcomes have varied considerably and have been inequitable. At the same time, the approach to achieving the outcomes and aspirations becomes much less centralized as the responsibility for curriculum, instruction, and assessment is left to local levels.

Standards-based reform leaves to those closest to students the decisions about which particular approaches to curriculum and instruction will best meet the goals of scientific literacy. Each professional will have to decide on the educational experiences appropriate for classes and individual students. Although admirable, this assumption presents a significant challenge for school personnel, especially science teachers.

A Change in Allocation of Funds and Resources

The aforementioned changes based on standards assume changes in the allocation of resources. Implementing standards for science education implies changes in facilities, equipment, materials, and in the end, budgets in order to specifically help students achieve the standards. Will resources be allocated to help all students achieve higher levels of scientific literacy? I certainly hope the answer is yes. Key indicators for such budgetary changes will include support for curriculum materials aligned with standards and support for professional development.

A Belief That All Students Can Learn

Although not necessarily an assumption unique to science education standards, it is nonetheless a fundamental assumption of standards-based reform that all students can learn science. Establishing the view that all students can learn science places equity in a central position in educational reform. This one assumption simultaneously brings honor to our profession and establishes the basis for significant and profound changes in programs and especially classroom practices. The commitment must be more than rhetoric and good intentions. If we assume that all students can learn science, then we must ask about attributions and opportunities when students do not learn. We will no longer blame students. Any attribution of blame must be shared by professionals in the education system. Just as we use the idea "justice for all" to indicate that all individuals regardless of race, gender, social stature, religious beliefs, and national origin must have basic legal rights such as due process, "science for all" implies that each and every student must be approached as an individual who can learn science and should be provided with appropriate and adequate opportunities to learn science.

The discussion in this section points to a number of implicit assumptions that should be explicit if we are going to realize the full potential of science education standards. At a minimum, it suggests that standards-based reform is more than a redefinition of content. More appropriately, the discussion highlights a number of issues that must be addressed as standards are implemented in programs and practices. Addressing the assumptions extends beyond the standards-setting process and includes issues of curriculum materials, instructional practices, assessment strategies, school governance, resource allocation, and most essentially, the professional development of science teachers. Very clearly, this discussion also assumed a systemic perspective.

Before proceeding to a discussion of the systemic portion of standards-based systemic reform, it is important to review some lessons from large-scale policy that apply directly to national standards for science education and indirectly to state and local standards.

Issues Related to Large-Scale Policies

Over four years of working on standards, I observed that content was the predominant interest of the majority of those who examined and reviewed the standards. I can extend this observation to the specific domains of physical, life, and Earth science subject matter. To be sure, the standards must have accurate and appropriate content. However, one must include other issues if the intention is to improve science education for the nation. There will have to be a reconsideration of commitments that include content *and* many other factors if we are going to successfully reform science education toward the goal of achieving scientific literacy. The issues of initiating and sustaining coordinated efforts to improve science education at the national, state, and local levels requires some attention to issues of scale, in particular, large-scale policy (Elmore, 1996).

Before discussing large-scale policy, it is important to address briefly the scale of the education system. Figure 9.1 and Table 9.1 provide two snapshots of the education system in the United States. This is a macro view of the education system; the numbers are approximate. This view provides a sense of scale when we talk about "the reform of science education" or "achieving scientific literacy for all students."

The scale of the policy efforts to achieve scientific literacy presents a critical, not so obvious, and seldom discussed, aspect of educational reform. A corollary to the general concern of large-scale policy issues is the translation of policies at national, state, and local levels, and the associated correspondence among the

FIGURE 9.1 *The United States Education System*

50 State Departments of Education
16,000 Districts (Superintendents)
2.8 Million Teachers
47 Million Students
115 Million Parents

TABLE 9.1 *The United States School System (ca. 1992)*

School District	16,399 district/ administrative offices 162,093 district administrators	449 county centers	174 Catholic dioceses	833 regional centers			
Elementary Schools	49,817 public elementary schools 1,083,096 teachers	3,397 public special ed schools 20,472 teachers	2,911 public combined K–12 schools 70,397 teachers	706 state–run schools	1,468 county schools 9,660 teachers	7,072 Catholic elementary schools 79,036 teachers	7,239 private elementary schools 50,984 teachers	4,604 private and Catholic combined K–12 schools 50,794 teachers
Middle/Junior High Schools	11,813 public middle/jr. high schools 346,523 teachers					115 Catholic middle/jr. high schools 898 teachers	83 private middle/jr. high schools 909 teachers	
Senior High Schools	14,010 public sr. high schools 532,844 teachers	1,288 public voc./tech. schools 23,973 teachers				1,173 Catholic sr. high schools 23,183 teachers	973 private sr. high schools 14,911 teachers	
Community College Four-Year Adult Education			635 adult education schools 10,214 teachers	1,566 two-year colleges 18,044 administrators 455,503 faculty	2,491 four-year colleges 41,726 administrators 455,503 faculty			
Other U.S. Schools	169 BIA schools 154 principals	397 APO/FPO schools 395 principals	1,785 U.S. territory public schools	686 U.S. territory private schools 421 principals				

Note: BIA = Bureau of Indian affairs; APO/FPO = army/fleet post offices.

policies. An understanding of large-scale policy is a valuable and almost entirely overlooked aspect of standards-based systemic reform of science education. To my knowledge, this issue has not been discussed in science education. In 1996 Richard Elmore discussed the issue of scale for educational reform in general. I draw upon the 1980 work by Paul Schulman, *Large-Scale Policy Making,* and my introduction to the idea through an article on climate change by Steven Rhodes (1992). I acknowledge here my reliance on their work for the following discussion.

The development and implementation of national standards constitute efforts of large-scale policy, and these efforts have characteristics that differ from other conventional policy-related efforts. Many government policies and programs are large in scope, numbers served, budget, and priority. National defense and education serve as examples. These, however, constitute normal government policies and programs. For Schulman (1980), the concept of large-scale policy involves an *extraordinary societal undertaking* rather than the conventional workings of government policies and programs. For contrast, the following discussion describes some characteristics of conventional policies: They are formulated and implemented in a piecemeal manner, they can change with elections and leadership, they can change with time to accommodate social conditions or to maintain the status quo. Think of examples such as tax laws, Medicare and Medicaid, agricultural subsidies, and in science education, Eisenhower funds. Schulman suggests that conventional public policy is "highly divisible." He means that the policies can be altered in discrete ways without endangering the sustainability of the policy. Small and incremental changes can be made through political tinkering without threatening the general policy.

In contrast to conventional public policy, consider policy initiatives that are comprehensive and systemic, ones that involve large-scale social aspirations. Concerning the nature of large-scale policy, Schulman (1980) states, "These are policy objectives whose pursuit requires comprehensive rather than incremental commitments and decisions. They require wide-ranging rather than piecemeal resource applications" (p. 15). He continues by elaborating other defining characteristics of large-scale policy: "Large-scale policy is defined by the existence of objectives that require, in their pursuit, the breach of certain thresholds. These thresholds generally [include] . . . technology, psychological receptivity, and political mobilization and support" (p. 16).

Discussion of several characteristics and examples of large-scale policy may support my contention that standards-based systemic reform of education should be thought of as large-scale rather than as conventional policy initiatives. Schulman provides analysis of three initiatives that he claims meet the criteria of large-scale policy: the space race with the Soviet Union, the war on poverty, and the war on cancer.

Large-Scale Aspirations

Each of these initiatives involved a major sociopolitical aspiration. With regard to the space race, to paraphrase President John F. Kennedy, we were to place a man on the moon and return him safely by the end of the decade.

In the contemporary educational reform, we also have a large-scale aspiration: to be first in the world in science and mathematics achievement. With this aspiration come attendant issues, not the least of which is how to achieve the goal and clarifying what it means (Hurd, 1992). As I have discussed, I do not think this is the goal on which science educators should focus their attention. I think goal three of *Goals 2000* is much more reasonable and admirable. It calls for higher levels of attainment at grades 4, 8, and 12, and it relies on standards to identify what students should know and be able to do (Bybee, 1996).

Public Support and Powerful Metaphors

The space race, the war on poverty, and the war on cancer were presented in patriotic terms in order to gain and sustain wide public support. Use of terms such as *race* and *war* engaged the public's sense of competition and conquest.

Suggesting "the nation was at risk" in the 1980s and that we required educational reform used a similar metaphor, and for the most part, it gained public attention. An additional three hundred or so reports proclaiming various aspects and remedies for the risk further supported the notion that we were at risk, primarily of economic decline, and that education would reinstate the United States to economic stability. Unfortunately, the metaphor of risk does not provide the same sense of action as *race* or *war.*

Long-Term Support and Definite Goals

Each of the large-scale policy initiatives required long-term and substantial commitments of resources. The space race had a goal—by the end of this decade (i.e., 1970); the war on poverty and the war on cancer did not have specific deadlines, but it was clear that they were too complex to be achieved in the short term and with small commitments of resources.

We are to be first in the world in science and mathematics achievement by the year 2000. In the world of politics and public memory, this represents a long-term goal for educational reform. When the goal was first announced, we had less than a decade to institute changes and achieve the goal. *Goals 2000: Educate America Act* set the time and the goals in 1994. Of course, by the mid-1990s, a Republican Congress was steadily dismantling those goals and the Democrats had other battles to fight. These political sources send inadequate and inappropriate messages to the educational community. This issue is also confounded by the fact that many educators have not realized the connection between goals and standards plus the role of standards as an educational innovation. In spite of the confusion, some educators have defended and supported them on professional grounds regardless of the political change.

Dependence on Experts

The three initiatives required the work of experts, largely those associated with engineering, social science, and medicine. The reliance on the scientific, engineering, and medical communities was most likely the result of those communities' being deeply associated with the war effort and subsequently moving into politics and public policy (Price, 1965). Curriculum reform in the 1950s and 1960s also was directed by the scientific community, as represented by, for example, Jerrold Zacharias, Glenn Seaborg, and Bentley Glass.

As discussed in prior chapters, the project to develop the *Standards* was located at the National Academy of Sciences and directed by scientists such as Jim Ebest, Richard Klausner, and Bruce Alberts. Further, scientists had considerable influence and input throughout the development, review, and final production of the *Standards*. Although this was appropriate and a significant contribution to the subject matter of physical, life, and Earth sciences, there were often conflicting perspectives on teaching, assessment, school science programs, and the principles and processes of educational change. In resolving such conflicts, we had to appeal to research literature in education.

Development of standards at the national, state, and local levels has involved various groups of experts, such as science teachers, science educators, and scientists. Their respective expertise provided strength and insight to the standards, but it also presented problems because the experts often did not agree and occasionally made recommendations in areas beyond their expertise.

Decline of Support and Independence

Eventually, the three major initiatives lost support and declined for different reasons. The space program lacked clear goals beyond reaching the moon. Probably, the war on poverty and the war on cancer declined because of the unexpected complexity of the problems and the difficulty of demonstrating progress toward the objective of solving them. As large-scale programs continue, it seems there must be tangible evidence of progress, or the public commitment wanes and eventually so does the financial support. With time, the three projects lost their independence, which contributed to further loss of support. It seems the lessons here reveal the role of accountability and sustainability: showing that the programs were indeed accomplishing what they set out to accomplish and that the personnel's original enthusiasm and public support were sustained over time.

It will be vital to demonstrate progress toward national, state, and local goals of achieving scientific literacy, or the public will lose interest and support will decline for standards-based reform. Educators should consider lessons from the space program, such as demonstrating the value of various projects after the initial Sputnik-spurred space race.

The insights about decline of support and independence also suggest the need to shift attention and efforts from standards to the changes required in programs and practices. This broader perspective on the science education system presents some hope of avoiding the problem of premature decline in the efforts to improve science education and achieve higher levels of scientific literacy. The counterpoint is the aforementioned, ever changing political winds.

Systemic Reform in Science Education

For purposes of study, scientists and engineers divide the natural world into segments, such as the solar system, the Earth system, the cell system, and smaller systems composed of elementary particles. The designed world consists of systems like jet airplanes, computers, calculators, safety pins, paper clips, and microchips. Social scientists study human organization, communication systems, transportation systems, and educational systems. A system consists of components that interact with each other and function as a whole. Some subsystems exist within other systems. One should clearly identify the boundaries of the system under study, for it is at the boundaries of systems that many important processes occur. In addition to system components and boundaries, there are flows of resources into systems, flows of products out of systems, and feedback mechanisms that adjust systemic processes.

We can view science education as a system with subsystems. For example, science classrooms have boundaries, and resources flow into the room in the form of textbooks and laboratory equipment. One assumes that the primary product that flows out of the classroom consists of students who have developed higher levels of scientific literacy. The science classroom exists within a district school system, which functions within the state education system, which functions within a national educational system. These examples point out an important idea: from a systems perspective, everything is connected to everything else.

The science education system is very complex. To develop a systems perspective of science education and the role of standards in it, we can look at some of the basic features of the science education system. What does the science education system produce? What does it do? The science education system should contribute to students' developing higher levels of scientific literacy. At the various levels, educators within the science education system set goals, establish policies and plans, develop curriculum materials, recommend teaching strategies and models, design assessments, administer programs, educate future teachers, and provide professional development for science teachers. All of these should serve the ultimate purpose of assisting science teachers in effecting scientific literacy of all students. Obviously, science educators must have a fairly clear conception of scientific literacy before the rest of the subsystems can be coordinated to serve that purpose. And the conception of scientific literacy must be translated from a general conception to more specific applications within the parameters of curriculum, teaching, assessment, grade levels, and teacher education.

What are the important components of the system? Components beyond science classrooms include teacher education, curriculum development, textbook publishing, educational technology, assessment groups, and administrative structures such as state departments of education and district administrations. The primary components within the boundaries of science classrooms include students, teachers, and the various materials and equipment that science teachers use.

What makes the science education system work? What are the actions or processes that influence the development of scientific literacy among students? Trying to answer these questions clarifies the complexities and realities of the systems perspective as it is applied to education. The most important processes focus on the interaction between science teachers and students, and on teaching and learning. Directly related to the interactive process of teaching and learning is assessing this process and designing experiences that optimize it. Designing experiences includes activities, such as the use of laboratory, curriculum materials, educational technologies, and field trips. An essential aspect is administrative support and budgeting to ensure that the science education system works.

What Do We Mean by Systemic Reform?

In the 1950s and 1960s, we characterized the reform as curricular. Most individuals describe contemporary reform as systemic (Cohen, 1995; Vinovskis, 1996; Banathy, 1992; Fuhrman, 1993; Fullan, 1996; National Science Foundation, 1996). But, what do we mean by systemic reform? Individuals use the word *systemic* in a variety of ways. *Systemic* can mean working with school systems and the structures of those systems. So, for example, some reform efforts focus on district finance and districtwide policies for science education. This view of systemic reform has a vertical orientation in that it attends to the reform initiative from top to bottom, and from bottom to top.

Systemic reform can also have a horizontal orientation, which means working with all schools in a district or state. So, for example, one would direct attention to science programs in all schools within a district and emphasize reform across the district. Another view includes both vertical and horizontal approaches to reform, takes a total systems approach, and initiates a variety of reform efforts. For instance, school personnel might begin by reforming the K–12 science curriculum and soon realize the importance of professional development, resource allocation, and administrative policies concerning assessment. The *Standards,* for instance, recognizes both the vertical and horizontal perspectives in the program and systems standards. Finally, some equate systemic reform with fundamental changes. In many respects, this view exceeds system reform in that its advocates assume the current system cannot change to accommodate contemporary demands, and so there is need to create new subsystems that may eventually develop the capacities of the system and thus replace the system. Support for private schools through tuition vouchers and charter schools provide two examples of this perspective.

The NSF (1996) document *The Learning Curve* provides a good definition of systemic reform:

> Systemic reform is a process of educational reform based on the premise that achieving excellence and equity requires alignment of critical activities and components. It is as much a change in infrastructure as in outcomes. Central elements include
>
> - high standards for learning expected of all students
>
> - alignment among all parts of the system: policies, practices, and accountability mechanisms

- a change in governance that includes greater school-site flexibility

- greater involvement of the public and the community

- a closer link between formal and informal learning experiences

- enhanced attention to professional development

- increased articulation between the precollege and post-secondary education institutions (p. 5)

Each of these views has some merit and some supporters. We have to decide on a systemic perspective. If we want to achieve scientific literacy, we need to shift our perceptions and efforts from a primary emphasis on events in the school system, that is, short-term efforts, to a focus on systemic structures that will provide enduring improvement of science education (Senge, 1990a, 1990b).

Reform in science education would be greatly enhanced by expanding perceptions and expectations from events alone to patterns of behavior and systemic structures. Most individuals in the science education community focus on events as a means to improve science education. For example, science teachers collect hands-on activities at workshops and implement those in classrooms. This perspective leads to an explanation of reform in terms of events, so "reform means using more hands-on lessons." Peter Senge suggests that some individuals view patterns of behavior as the focus of change. This view takes a longer-term perspective. Examples for science education include those who recognize that reform includes the long-term changes in teacher education, both preservice and continuing professional development. Such views extend (and displace) the immediate and reactive views of event-based approaches. The most powerful view, because it helps explain the other two, is structural or systemic. Systems explanations address the underlying causes for events and patterns of behavior, so they imply a means of change and improvement. Understanding scientific inquiry can change the patterns of teaching behavior and activities in ways that are more significant and enduring than merely supplying teachers with new activities. This example points out the generative nature of systemic change. The inquiry example also shows that improvement relates directly to how individuals think about science education.

Some Aspects of Systems Thinking

One of the central indicators of systems thinking is being able to see the big picture—the forest *and* the trees at the same time. The big picture refers to the size, timing, and components of the system. For example, a science supervisor would recognize the number of teachers and students in the school system, the time it takes to change (multiples of years), and that curriculum, assessments, administration, and community must all be included in the change process.

Another aspect of systems thinking is being able to recognize interrelationships and processes rather than single components and products. How many times have you heard an educator express the view that some single factor is *the cause* of various school problems? I have observed that teacher preparation and assessment are most

TABLE 9.2 *Recognizing Standards-Based Systemic Change in Science Education: Indicators of Progress*

Perspectives of Change	Maintenance of Traditional System for Science Education	Initial Awareness	Transition of Science Education	Predominance of New System for Science Education
Purpose:				
Goals	Support for old visions	Discussion of new goals	New vision clarified by stakeholders	New vision in place
Vision	Demeaning statements about science literacy	Recognition of new ideas	Recognition of connections in system	Support for purpose and goals statements
	General resistance to change		Exploration of goals statements	
	Appeal to past			
Policy:				
Frameworks	Use of old policies, frameworks, syllabi	Discussions of committees for new frameworks	Committees form frameworks for curriculum, instruction, assessment	New frameworks for curriculum
Syllabi	Same budget priorities	Public discussions of change	Beginning of resource allocation	Budgets fully reflect new vision
Budgets		Budget considerations	Standards (outcomes) defined	Standards in place
Standards		Recognition that policies need review	Discussions of standards	

TABLE 9.2 Recognizing Standards-Based Systemic Change in Science Education: Indicators of Progress (continued)

Perspectives of Change	Maintenance of Traditional System for Science Education	Initial Awareness	Transition of Science Education	Predominance of New System for Science Education
Program:				
Curriculum	Use of traditional programs	Recognize need for new programs	Review of innovative curriculum	Implementation of new curriculum
Assessment (school or state)	Dated textbooks, equipment, materials	Need for curriculum reform	New assessments explored	Multiple means of assessment
	Standardized testing		Professional development proposed	Administrative support for new program
			Nation, state, local policies aligned	Professional development aligned with new program
Practices:				
Teaching Strategies	Resistance to innovative practices by majority of teachers	Recognize disparity between current teaching and research	Workshops on new approaches	Teaching using active learning
Assessment (classroom)	Delivery of facts and information	Awareness of new emphasis	Planning time for teachers	Resistant teachers in minority
	Few activities	Discussions of new approaches	New modes of assessment implemented	

often mentioned, especially if one is discussing curriculum and instruction. Other examples include the administrators or parents, or the school board, as *the cause* for the lack of innovative teaching. Such views consider each component of the science education system as an isolated event or causal factor. Somehow, those components are not perceived as being interrelated, as parts of continuing processes that occur in the education system.

Those who exercise systems thinking spend less time on blame setting and more on problem solving. Many individuals devote excessive time to identifying and attributing blame, almost always to outside circumstances or people other than themselves. Systems thinkers recognize that the causes (and solutions) for problems are, by and large, not attributable to individuals. The problems generally lie within the system.

Some individuals focus on details when it would be much more productive to focus on dynamics. This is especially true for complex issues. Attending to details alone is akin to treating symptoms but not the disease. The dynamics of complex problems are often remote in time and space, and the interventions are often not obvious.

Systems thinkers have the ability to find the core of problems and are thus able to implement small, well-focused solutions that have broader, long-lasting impact. In science education reform, one would try to identify difficult problems at the points of highest leverage, where a minimal change would result in enduring widespread improvement. Many individuals recognize this, at least intuitively, when they suggest changes in assessment, curriculum, or teacher education.

One final aspect of systems thinking is the ability to recognize the underlying causes of problems rather than just the symptoms. Unfortunately, the pressure to intervene and *do something* in school systems is quite high. The result is endless workshops on symptoms. But the problems return. Whether in the science class-room, school district, or state, one can see the benefit of working on the causes of problems and initiating improvements at that level rather than expending time, effort, and money alleviating symptoms.

Stages of Systemic Change

In this period of standards-based systemic reform, one can become overwhelmed with the multiple competing forces for change. Knowing where to begin or recognizing patterns of change is often difficult because of the different perspectives: purpose, policy, program, and practice, and the various stages of reform that a state, district, or school may be in.

Table 9.2 presents different perspectives and indicators to help teachers, administrators, parents, and others in the science education community gain some understanding of the process of systemic reform. The framework in the table uses the perspectives of purpose, policy, program, and practice, and some indicators of progress in systemic improvement. The idea for the framework came from several sources (Anderson, 1993; Fullan, 1993; Hall and Hord, 1987; Loucks-Horsley, 1995; Loucks-Horsley and Stiegelbauer, 1991; Fullan and Stiegelbauer, 1991). The framework will be useful to develop strategic plans for improving science education, evaluating progress in reform, and assessing the outcomes for different aspects (perspectives) of reform.

This framework only provides a general view of systemic change. Differences will exist at national, state, and local levels. Various perspectives will differ, for example, purpose statements usually precede policies and programs.

Conclusion

Standards for science education provide the foundation upon which to improve curriculum, instruction, assessment, and professional development. Standards serve different functions in the system we call science education. They are a call to action, a measure of what we value, a definition of scientific literacy, and feedback for the reform of different components of science education.

National standards represent large-scale policies that hold great potential. Assumptions underlying the standards will challenge all those who have anything to do with improving programs and practices in science education. Truly implementing the standards will vastly change the system.

The promise of accomplishing the vision and aspirations of the standards lies in the systemic nature of the contemporary reform. Although comprehensive and complex, the systemic nature of this reform implies that all components must contribute to the purpose of achieving scientific literacy. The burden of reform will not be placed on components such as curriculum materials, as in the 1960s and 1970s. Rather, leadership across the entire science education community will share the burden.

10   *The Year 2000 and Beyond*

The goal of achieving scientific literacy for all students presents a complex set of challenges. The science education community has the knowledge, values, and abilities to address the challenges and implement the required changes. This chapter discusses a variety of issues associated with those challenges, the themes of this book, and the particular goal of achieving scientific literacy.

The Myth of Scientific Literacy

In his book *The Myth of Scientific Literacy* (1995), Morris Shamos presents a view that seems counter to mine. He claims that scientific literacy is for all intents and purposes an unachievable goal and that it has led astray all those who are trying to improve science education. Here is a brief summary statement from his book: "By now it should be apparent that the notion of developing a significant scientific literacy in the general public, as we have come to understand its normal meaning, is little more than a romantic idea, a dream that has little bearing on reality" (p. 215). Shamos concludes that we might consider a *new* sort of scientific literacy, namely, scientific awareness. The guiding principles for presenting science to the general (nonscience) student should focus on teaching science to develop an appreciation and awareness of the enterprise, i.e., a cultural perspective; using technology as a central theme, i.e., a practical perspective; and emphasizing the proper use of scientific experts in science-related social problems, i.e., civic science literacy. I believe that even though he dismisses scientific literacy as an unachievable goal, Shamos' own goal of scientific awareness is not inconsistent with those of others (e.g., Shen, 1975) who have discussed scientific literacy.

The difference seems to turn on the interpretation of the word *myth*. Shamos uses the term in its meaning of "half truth" or "falsehood." But from a historical or classical perspective, myths are stories that provide explanations supporting a particular worldview. They establish bridges between the known and the unknown.

208

In short, they help individuals make sense of the world. In this sense of *myth,* I agree with Shamos that scientific literacy is a myth.

Myths help individuals define and clarify issues that are within our immediate understanding but not evident. Myths serve the important purpose of supporting a worldview; they appeal to the consciousness of a people by embodying explanations, aspirations, and ideals of the culture because they express deep commonly felt emotions. The late Rollo May discussed the importance of myths in one of his last books, *The Cry for Myth* (1991).

I believe scientific literacy in this meaning of *myth* serves an important purpose for the community of science education. It helps us express our goals and aspirations for science education; it brings meaning to our work and makes connections between diverse individuals in our community. Perhaps complete scientific literacy is unachievable and stands as an unrealistic expectation. But almost any way of expressing our purposes would be subject to the same criticism.

Think of some of the great myths: the self-discovery of Oedipus, the tragedy of King Lear, the American dream and the tragedy of Jay Gatsby, the individualism of Horatio Alger, the confrontation with evil in *Moby Dick,* and the continual struggles of Sisyphus. Mythical perspectives provide a larger, more holistic view that unites many specific features such as the historical and the present, the individual and the social. And they convey something that those with our contemporary interest in constructivism should embrace; namely, they have meaning and significance at a personal level. So too does scientific literacy for the science education community.

The myth of scientific literacy contributes to our professional life in several ways. It gives a sense of identity. Discussing, even arguing against, scientific literacy answers questions about who I am professionally. It also provides a sense of community: it gives all of us in the science education community a common purpose for bonding and a deeply rooted connection with colleagues.

The myth of scientific literacy has an essential function in science education. It is like the foundation for a house. Generally, foundations are not exposed to outside view, but they are essential and they provide the support and structure on which to build. It is the case that, once set, the foundation determines many features of the house. Still, other individuals such as contractors, carpenters, and residents make their unique additions and modifications to the house.

If scientific literacy is the foundation, *National Science Education Standards* (National Research Council, 1996a) provides the specifications and framework for the structure. This said, it should be clear that many features remain to be completed by other professionals. I refer to these as the programs and practices of science education.

My own view, and the one that is the theme of this book, is that achieving scientific literacy serves the purpose of a goal, one that accommodates changes in science, society, and subsequently science education. We have progressively clarified the goal and continually endeavored to achieve it. Like many goals, it must be understood and implemented in a contemporary context. As a society we are well served by purposes such as "justice for all." When we do not achieve this goal, what is the proper recourse—to change the goal or to refocus our efforts on those aspects of the goal that we are not achieving? I think that we would maintain the goal and reform our efforts. Achieving scientific literacy for all students should remain our purpose while our efforts within the educational community evolve to better achieve that purpose. However, we must face certain contemporary realities.

The Reality of Achieving Scientific Literacy

The reality of achieving scientific literacy rests on our ability to bring about changes in all components of the system, and to attend to what ultimately counts— science teaching and student learning. Here, the science education community must face a reality of history: in the United States, numerous and varied reform efforts had little effect on teaching and learning in classrooms. The evidence seems clear. Educators talk of new goals, institute new slogans, and change curriculum frameworks, but teaching and learning in classrooms continue with little change. These issues are confounded by attacks on the system (Berliner and Biddle, 1995). Richard Elmore and Milbray McLaughlin (1988) expressed the fundamental issue quite insightfully when they said, "Reforms that deal with the fundamental stuff of education—teaching and learning— seem to have weak, transitory, and ephemeral effects; while those that expand, solidify, and entrench school bureaucracy seem to have strong, enduring, and concrete effects" (p. v).

The insight that I derive from this is that we must identify the means to change the fundamental stuff of teaching and learning so they are strong, enduring, and have concrete effects. Specifically, we need fundamental changes so that science teachers are more effective and students achieve higher levels of scientific literacy.

Richard Elmore further clarified the issue of fundamental change in a 1996 article, "Getting to Scale with Good Education Practice." He summarized the problem of scale in educational innovations as follows:

> Innovations that require large changes in the core of educational practice seldom penetrate more than a small fraction of U.S. schools and classrooms, and seldom last for very long when they do. By "the core of educational practice," I mean how teachers understand the nature of knowledge and the students' role in learning, and how these ideas about knowledge and learning are manifested in teaching and classwork. The "core" also includes structural arrangements of schools, such as physical layouts of classrooms, student grouping practices, teachers' responsibilities for groups of students, and relations among teachers in their work with students, as well as processes for assessing student learning and communicating it to students, parents, administrators, and other interested parties. (p. 2)

The reality of achieving scientific literacy means that science educators must address issues at the educational core and recognize that tinkering at the edges will not suffice. The problem of scale can certainly be viewed as a set of interrelated problems regarding purposes, policies, programs, and practices all situated at different places in the educational system. Further, the problem of scale is not solely one of change. States, schools, and science teachers change all the time. The problem seems to be that we change peripheral things and do not change the core.

How do we get from the goal of achieving scientific literacy to the fundamental stuff or core of science teaching and student learning? My answer lies in the theme and framework of this book. One can conceptualize the reform of science education as operating in four connected and interdependent areas: purpose, policy, programs, and practice. Although each area has its own advocates and leaders, audiences, problems, and perceptions of how the educational system operates and what is important within the system, it is also important for the science education community to recognize the common purpose and then strive to reduce the variance among them.

The Purpose of Teaching

Science educators express aims, goals, and objectives in various documents such as national standards, state frameworks, school syllabi, and teaching lessons (DeBoer, 1991; Bybee, 1993; Bybee and DeBoer, 1993; DeBoer and Bybee, 1995). The term *purpose* refers to universal goal statements of what science teaching should achieve. Such statements are abstract and apply to all concerned with science education. Achieving scientific literacy is a contemporary statement of the purpose for science teaching. The strength of this purpose lies in its widespread acceptance and agreement among science educators. The weakness is its ambiguity about specific situations in science education and what it means to achieve scientific literacy for individual students and for the entire population of students. For example, what does the goal of achieving scientific literacy mean for a grade 3 teacher? a high school chemistry teacher? a teacher educator? a policy maker? a curriculum developer? The answers, of course, vary for different situations—hence the need for more concrete statements of scientific literacy and the adaptation of purpose statements for various components of the science education community. Statements that are based on the purpose of achieving scientific literacy but which address more specific components of the educational system introduce the role of policies.

Policies to Achieve Scientific Literacy

Policy statements are concrete translations of the purpose for various components of the science education community. Documents that give direction and guidance, but are not actual programs, define policies. Examples of policy documents include course plans for high school science classes, district syllabi for K–12 science, state frameworks, and national standards. Similarly, college or university requirements for undergraduate teacher education and state and national frameworks for assessing scientific literacy fall into the category of policies for science educators. In the United States, major policy documents include the *National Science Education*

Standards (National Research Council, 1996a) and *Benchmarks for Science Literacy* (American Association for the Advancement of Science, 1993). If we are to achieve higher levels of scientific literacy, they must directly and consistently inform decisions at state and local levels.

Programs for Science Education

Science programs include the actual curriculum materials, textbooks, and courseware based on the policies. Programs are unique to grade levels, disciplines, and aspects of science education, such as teacher education or secondary school science programs.

School science programs may be developed by independent national organizations, state departments of education, and commercial publishers, or they may be developed by local school districts. Who develops the materials is not the defining characteristic; the fact that schools, colleges, state agencies, and national organizations have programs aligned with policies such as the national standards is the important feature of this aspect of reform and an essential link between achieving scientific literacy and improving science teaching.

Practices of Science Education

Practice here refers to the specific actions and processes of teaching science in schools, colleges, or universities. The practices of science education include the personal interactions between teachers and students and among students as well as the roles and uses of assessment, educational technologies, laboratories, and other methods of teaching science. In the framework described here, implementing new classroom practices implies they would be consistent with policies, and school science programs would be designed to enhance learning and enable students to achieve higher levels of scientific literacy.

Improving the practices of teaching science focuses on the individual, unique, and fundamental aspects of education. Science educators can propose new goals and design new standards, syllabi, and scope and sequence charts. And they can develop new curriculum materials. But the critical aspect of the contemporary reform is improving teaching and learning in classrooms. Concerning classroom practices of science teachers, it is clear to most educators that practice can reflect effective programs, policies, and purposes, but they are not always consistent. In fact, it is quite the opposite. In general, the science education system—policies, programs, and practices—can be characterized as incoherent; that is, they have considerable inconsistency and variation. Common sense and reason suggest that the *Standards* could bring greater consistency to programs and practices.

Using this framework (the Four Ps), one can imagine educational reform originating and developing in four fundamental ways:

- Professional science teachers change their perceptions of, and strategies for, more effective practice.

- School administrators and curriculum developers change the school science program.

- Policy makers change the authoritative documents, such as standards, that identify conditions for effective programs and practices.

- Prominent educators provide statements of new priorities and emphasis for educational goals.

Although any reform has a historical order, there is not necessarily a logical or sequential order implied in my discussion. To be specific, my discussion of the Four P framework does not have a top-down, hegemonic perspective. In fact, reform can originate from different workers in the educational system, for instance, teachers, policy makers, curriculum developers, or philosophers. The truth is that contemporary reform has generally progressed from purpose; that is, the 1989 Governors' Summit defining educational goals and by implication identifying scientific literacy for science educators, which then influenced policies such as the *Standards*. Now it seems reasonable to argue that our concern should focus on programs and practices. I would point out the historical fact that the reform of the late 1950s and 1960s began with the reform of curriculum, the general domain that I refer to here as program. That reform proceeded to recognize the need to change practices vis à vis academic year, summer, and inservice institutes supported by the NSF.

A systemic view suggests that there is (or should be) an interdependence and coherence among practice, program, policy, and purpose. Further, there is probably not simultaneous reform in all domains. However, if there are clear signals and consistent information throughout the educational system, one can imagine a progressive alignment over time.

This said, it seems evident that there is greater coherence and consistency within a domain, such as policy, than between domains, such as policy and program and practice. In examining state frameworks and using the *Standards* and *Benchmarks* for the standard of comparison, one is struck by the inconsistencies that enter the system. The inconsistencies become even more pronounced as one examines local frameworks (Council of Chief State School Officers, 1995b). It seems that rewards, incentives, and resources tend not to emphasize the translation of policies to programs and purposes to practices. Policy makers also tend mostly to communicate with other policy makers, and practitioners tend mostly to communicate with other practitioners. If we are truly interested in widespread, large-scale, and long-term educational reform, considerable time and effort must focus on the interface between the respective domains, and all those in the science education community must recognize that reform will occur through coordinated, consistent, and coherent translation of purposes, to policies, programs, and practices.

This analysis returns us to the importance of science teaching and learning in achieving scientific literacy and the premise that reform must be grounded in the agreement, clarification, and wide-spread commitment to what science content all students should learn, what teaching strategies will consistently enhance learning, how assessments should be aligned with the valued science content, how the school organization and culture select for these changes, and what science teachers will require in terms of professional development. This line of discussion leads to the essential role of standards and the links between standards and teaching and learning.

The Potential Effect of Standards on the Core of Educational Practice

Since the early 1980s, the pressure to improve science education has increased in the United States. In the period from 1980 to 1996 the response to this pressure has generally become more confused and incoherent within the educational system. For example, many states mandate increased graduation requirements without changing science content, or states institute testing systems without changing guidelines for textbook adoption, or they implement new curriculum without changing tests or providing professional development consistent with innovations in programs and practices.

The reason for the variations among states and communities lies in our history and in the rights and responsibilities granted to states and local jurisdictions by the Constitution. Although states have considerably more authority than federal agencies or national organizations over education, the power to determine the curriculum, instruction, and assessment is usually delegated to local school boards and ultimately to classroom teachers.

In 1989 the Mathematical Sciences Education Board of the NRC presented the results of a study that sought to explain differences between the mathematics curriculum in the United States and other countries. The study was motivated by a need to explain differences in student achievement on international assessments. The following summary from *Everybody Counts* (National Research Council, 1989) presents the situation for mathematics, and one that most would agree is similar for science education:

> In the United States, with our traditional and legal decentralization of education, we go about things very differently [than other countries]. Every summer, thousands of teachers work in small teams for periods ranging from one week to two months, charged by their school districts to write new mathematics curricula. These teacher teams usually have little training in the complicated process of curricular development, little or no help in coping with changing needs, and little to fall back on except existing textbooks, familiar programs, and tradition. The consequence usually is the unquestioned acceptance of what already exists as the main body of the new curriculum, together with a little tinkering around the edges. Many school districts simply adopt series of textbooks as the curriculum, making no effort to engage the staff in rethinking curricula; in those places, the status quo certainly reigns.

> The American process of curricular reform might be described as a weak form of grassroots approach. The record shows that this system does not work. It is not our

teachers who are at fault. In fact, teachers should play a dominant role in curriculum decision making. But teachers who work in summer curriculum projects are being given an unrealistic task in an impossible time frame, with only the familiar status quo to guide them.

In static times, in periods of unchanging demands, perhaps our grassroots efforts would suffice to keep the curriculum current. In today's climate, in which technology and research are causing unprecedented change in the central methods and applications of mathematics, present U.S. practice is totally inadequate. (pp. 77–78)

It is time to acknowledge the fact that this grassroots approach is inadequate to our contemporary challenges. Science education is left with a system that lacks curriculum consistency, instructional coherence, and assessments congruent with our purposes. Using the *Standards* can bring greater consistency to the system by clarifying what teachers are to teach and what learners will be expected to know and be able to do. Further, assessment would be aligned with the valued outcomes achieving scientific literacy. Grassroots improvement now means writing of assessments, improving instruction, and implementing coherent professional development programs.

This line of discussion leads to several conclusions. First, gaining widespread acceptance and agreement on what science teachers should teach and what students should learn is required. Second, the need for greater consistency and coherence among the educational system's stated purposes, policies, programs, and practices is essential. Third, much greater emphasis should be placed on the work of those who translate policies to programs and implement programs and practices. Finally, the greater the freedom of professionals to design their own programs and practices, the higher is the need for enlightened professionals. Here I am using the term *enlightened* to mean possessing greater knowledge, understanding, and abilities associated with both science and education.

What is the evidence that implementing standards can or will make a difference in science teaching and student learning? Most discussions of standards rely on some combination of common sense, reason, and intuition to support the position that implementing standards will improve teaching and learning. Although these are not justifications that can be ignored, it may be important to look beyond such justifications for any evidence that standards make a difference. I note here that the implementation of specific standards, such as the *National Science Education Standards* (National Research Council, 1996a), is in the early stages in the United States, so little evidence is available. However, many other countries that we generally compare ourselves with do have standards (or other mechanisms that function like standards, e.g., national curricula). This discussion relies on evidence from international comparisons such as those in *The Learning Gap* (Stevenson and Stigler, 1992), the 1996 Organization for Economic Co-operation and Development report *Changing the Subject* (Black and Atkin, 1996), and results from the TIMSS (National Center for Educational Statistics, 1996).

In the United States, most attention to the international comparisons has focused on results that report low achievement levels for our students. International comparisons have generated healthy debate among educational researchers, who tend to address methodological problems (e.g., Bracey, 1996), and social scientists, who seek explanations for the variance in social factors such as number of parents at home, parental education, economic status, and type of community. Some discussions extend to the influences of social psychology, such as perceptions of the importance of ability versus effort in student achievement (Stevenson and Stigler, 1992). No doubt, all of those factors influence student learning and even may be significant explanations for the variance in achievement. However, schools have little influence over such factors as parental income and marital status. Schools do have an influence on standards they set for all students. And it seems reasonable to look at any evidence supporting the potential effect of standards on teaching and learning. Here, we look to international comparisons and ask if in other countries national standards have an effect on teaching and learning. The answer seems to be yes. A general review of countries that do better than the United States indicates that most of those countries have standards (although they are often not referred to as standards) that are consistently applied. Further, those countries have means of implementing programs that are consistent with those standards and using assessments that are congruent with standards (National Center for Educational Statistics, 1996; Schmidt, McKnight, and Raizen, 1997).

Considering different educational systems, in particular those that do well on international assessments, one can ask, What factors seem to be important for teaching and learning? Many countries, such as Japan, have a central ministry that asserts what children should learn in order to become good and productive citizens. They further know how to organize experiences that will promote learning. A central ministry of education ensures that all students will be exposed to the same curriculum, be taught by teachers using similar approaches, and will be expected to attain the same general standards of performance.

The positive effects that standards have on teaching and learning seem to arise from several fundamental components of any education system. I summarize here the guiding principles for those components.

Consistent Curriculum

Although this may seem obvious to those who review and think about curriculum, it is not the case that all U.S. school science curricula are consistent with national goals such as the *Standards*. This is generally true within this nation, the states, and local school districts. For example, the textbooks may vary, the emphasis in different states or schools may vary, and the actual topics may vary. The model implied by the *Standards* is one of consistent conceptual and procedural goals across the nation, states, and local school districts. That is, the design of curriculum materials should enhance the learning of content described in the *Standards*. Variations could well be found in the topics, activities, and sequences. However, the conceptual organizers, and fundamental concepts within grades or age levels, would be consistent within the system. I note that standards-based curriculum does not imply a national curriculum any more than the present textbook approach results in a de facto national curriculum. State and local school districts have considerable latitude in the selection, organization, and emphasis of topics and activities.

Coherent Instruction

As Harold Stevenson and Jim Stigler (1992) point out, one way to think of instruction is to use the analogy of a story. Just as good stories are organized and have a beginning, middle, and end, so too should science lessons. We can take that a step further. Stories first engage the reader's interest and then proceed to develop themes through a series of interconnected and interrelated events. It is essential that instruction—the science lessons—be carefully designed to engage the learner and slowly develop concepts or abilities through a series of experiences. My point stresses systematic and coherent instruction and not a particular model. In Chapter 8, I did provide one model that has been used in a variety of science programs. Although I certainly encourage the implementation of this model as a means of bringing greater coherence to instruction, there are other models that will achieve the aim of coherent instruction.

Congruent Assessment

Assessments should be congruent with the valued objectives of the curriculum. Again, this may sound obvious, but in many instances tests and assessments are not congruent with other components in the educational system. The principle is simple: the form and function of assessment should coincide with the form and function of science content outlined in the standards.

Continuous Professional Development

The Japanese provide a great lesson on the effect of standards on teaching and learning. Namely, the continual work of teachers includes improving lessons, "polishing the stone," through the provision of time and opportunity for professional development. Professional development of science teachers will be one of the critical variables in achieving scientific literacy. It is too much to expect teachers to change without the emphasis and support provided through continuous professional development.

Challenging Our Current Models

This section presents one analysis of contemporary reform. I answer two interrelated and fundamental questions: What needs to change? and How should we approach the improvement of science education? This analysis presents a critical response to current perspectives and recommends new perspectives.

The Use of Science as a Metaphor for Reform

I was struck by the pervasive use of technological metaphors in Kenneth Wilson and Bennett Davis's book *Redesigning Education* (1994). Although the authors argue for a paradigm shift à la Thomas Kuhn and provide scientific examples such as the Copernican Revolution, they clearly present technological metaphors and examples as the means to achieve the new educational paradigm. For example, the agricultural extension system, the redesign process, market resistance, and the lack of a technical culture. Note that the book's title even expresses a technological solution. Perhaps we should acknowledge the limited influence of developing scientific explanations for various educational issues and recognize the design process and problem solving as the effective means of improving science education.

One often hears requests for "the evidence" that changing the science curriculum or instructional practices will indeed achieve the goal of scientific literacy. Or, will developing higher levels of scientific literacy really result in higher levels of economic productivity for the nation? Note that individuals asking these questions have a perspective grounded in a metaphor of science. Namely, evidence leads to better explanations and understanding; or, with the common technological extension, scientific knowledge can be applied to solve practical problems. Scientific evidence for issues related to science education certainly seems appropriate, and it is quite understandable for those who practice, teach, and educate about science to have a worldview and mental model based in science. However, I suggest that using a scientific metaphor may not be the most productive approach to educational reform. Technology or design may be a better way to think about reforming and improving science education programs and practices.

I am making a distinction between science and technology that assumes science originates in questions and uses evidence to propose explanations or answers for those questions. Technology, on the other hand, begins with a problem and attempts to solve the problem using whatever materials, processes, or knowledge are available. To be sure, technological solutions rely on scientific understanding and must accommodate scientific principles. Technological problem solving also includes other factors such as criteria for the solution, constraints such as assessing cost, risks, and benefits, and making trade-offs in order to finally solve the problem. As one considers every step in the process of translating a purpose such as achieving scientific literacy, there are criteria, constraints, and trade-offs. Science teachers present these all the time: The materials have to work with my students. What will we do about college-bound students? Does the new program require computers? If so, we can't adopt it because we don't have them.

In order to achieve higher levels of scientific literacy we need to move beyond an exclusive emphasis on explanations of curriculum, instruction, and assessment. We have these. We need new designs and new solutions to problems. Returning to a metaphor used in the discussion on myths, we need individuals who can design new foundations. And the first step in the design process is identifying the purpose and core values that should guide the overall process. Of course, I nominate the purpose of achieving higher levels of scientific literacy for all students. If this is the first task, what is the second? The second is identifying the policies, the strategies, the guides that will inform decisions about more concrete issues of programs and practices. The *Standards* embodies this purpose. They are the next step in the process of designing improved educational systems. The third and fourth tasks involve developing programs and implementing practices that are consistent with the *Standards*.

Behind this entire design process there must be enlightened professionals, and that goal requires professional development. Educational systems need more than a lofty purpose and guiding policies. The design requires a process of continuous improvement by those who have the responsibility for teaching and learning.

We should recognize the various problems of reform and try to solve them head-on rather than providing explanations or seeking evidence to support our personal mental models. I will underscore the following point. Switching our dominant metaphors from science to technology does *not* eliminate the need and importance for research and evidence for the efficacy of various reform initiatives. It does, however, suggest a more applied approach to research in science education and thus greater emphasis on the long-standing void between theory and practice.

The Overuse of Innovations

Education in general, and science education in particular, can easily be faulted for the overuse of innovations and the underuse of fundamental changes. Rather than actually implementing a basic reform such as teaching science by inquiry, we have sought resolution for all problems in single innovations. Constructivism and teaching for conceptual change currently qualify, as does cooperative learning, for the innovations that will lead us to successful reform. Let me be clear about this. It is not the innovation that I am criticizing; it is our *overuse* of the innovation that presents problems. The probability of any single innovation improving the entire system of science education is quite low.

Overusing single innovations reveals a lack of a vision, an inadequate perspective of systemic reform, and an avoidance of fundamental solutions. We simply must begin to recognize that most innovations have an appropriate place in science education, and we will all have to make decisions about the best ways and means for implementing the respective innovations.

The Avoidance of Fundamental Solutions

For about two decades, we have provided easy solutions to problems such as teaching science by inquiry. We reduced the process of inquiry to "the processes" and implemented the solution in the form of activities that required little or no understanding of either inquiry or science and could be used "on Monday." Teachers of science are now so used to "make-it-take-it" workshops for hands-on materials that we provide almost nothing else at our inservice meetings and at state, regional, and national conventions.

At one time providing lessons to use tomorrow seemed like a good solution, but now the solution has become a significant part of the problem. Such approaches to reform and improvement of science education do not represent a systemic approach, and they clearly avoid a solution to fundamental problems. There is, however, an even more detrimental result of avoiding fundamental solutions. Providing short-term solutions is not just ineffective; it is addictive in the sense that it results in an unintended dependence. Science teachers require hands-on workshops and have little tolerance for any suggestion of other approaches. Such a position reduces their ability to solve their own problems through the design and development or adoption and adaptation of science programs.

The science education system is fundamentally weaker than before the extensive provision of practical solutions in the form of hands-on, how-to lessons. There is a significant need to help individuals in all components of the system assume responsibility for solving their own problems.

One of the first questions about the *Standards* is, Where is the curriculum that meets the *Standards?* The unfortunate perspective that underlies this question is that the individual does not want or need to understand the *Standards,* only to implement the appropriate program. I think that responding to these immediate needs countervenes initiatives to empower school personnel. When they abdicate their responsibilities, teachers give up the power gained through knowledge and understanding of the *Standards* and the focus on fundamental solutions. This is not to say that they will have to assume responsibility for the entire implementation process. It does imply that they will have to understand their purposes and make informed decisions about various aspects of programs and practices.

The Erosion of Fundamental Goals

For at least two decades we have continually reduced our goals and expectations. We have "passed" students who do not understand science, and we have "dumbed down" science textbooks and lowered graduation requirements. Science educators at all levels have lowered their expectations and standards, or allowed them to be lowered by administrators, parents, or publishers. When we have not explicitly lowered expectations for students, we have implicitly lowered the quality of our science programs by cutting budgets, increasing class size, and needlessly implementing the latest fad or innovation.

While we reduced our expectations we lost sight of the fundamental goal of developing literacy for all students. The standards-based reform is an explicit attempt to raise expectations and refocus our attention on fundamental goals.

The Emphasis on Individual Responses

We have all tended to avoid our own responsibility for improving science education by looking to others for solutions. Whether it is national standards, a new science program, or a professional development workshop in a school district, we place the burden for the entire reform on the individual or project. Seldom have I heard anyone say, "BSCS Middle School Science and Technology is a good middle school curriculum; I will have to modify my teacher education program." Comments usually take the form, "What will you do about teacher education? Changing teacher education is essential if you want to improve science education in middle schools." My point is this: We have to avoid the tendency to place blame on, or shift the burden of responsibility to, others.

What we really need is, first, a systemic view of reform, and second, the realization that we are all part of the system. Thus, in some measure we are all responsible for the reform.

The Intense Use of Common Resources

Science educators may recognize this as a "tragedy of the commons" problem. Individuals, usually school boards, superintendents, and administrators, continually intensify the use of common resources, especially teachers. Examples include lunchroom, hall, and bathroom duty, enlarging class size, increasing the diversity of classes, and adding topics to the curriculum without eliminating any. Like other solutions, this one works for awhile, but eventually there are diminished returns. How can you tell when the human resources reach their limits? There are increasing complaints about conditions, more teachers teaching out of their discipline, the loss of enthusiasm and motivation for teaching science, divisive relations between teachers and administrators, decline in morale, burnout, and steady attrition of the workforce.

The Escalation of Changes in Science Textbooks

Escalation occurs when two or more people or organizations, who see their welfare as depending on the relative advantage one has over the other, continually act to gain advantage by reacting to advances by the other party. Science textbooks are subject to the problem of escalating changes. In order to capture market share, publishers have responded to adoption requirements in major states, increased the size of books, accelerated the revision cycle, added to the package of materials, given more free samples, and increased marketing strategies. All of this has increased the cost to publishers, decreased the diversity among programs, and emphasized vocabulary and technical terms—the lowest common denominator of scientific literacy.

The escalation of changes in textbooks brings forward several models that we must challenge. The first is state adoptions. In a climate of standards-based reform there is a real need for adoption requirements to be consistent with national and state standards. Although the majority of states are changing frameworks (Council of Chief State School Officers, 1995), there must be corollary changes in adoption requirements. Second, there is the cost of attending to all the requirements of different states. The cost is not only financial but also intellectual and educational. Most emphasis is placed on surface features such as topics and table of contents, and least emphasis is on the opportunities and requirements for learning. Relative to the *Standards,* alignment with the *Standards* consists mainly in aligning subject matter topics and not other features such as the broad range of content, e.g., inquiry, history and nature of science, technology, and personal and social perspectives in science. There is little alignment with respect to the standards for teaching, assessment, and professional development. Third, and related to adoption requirements, is the cost of providing sample copies of textbooks and accessories such as overhead transparencies and other supplementary materials. Finally, the cost of marketing must have increased, because the differences among programs is insignificant, so greater efforts and finances must be expended in order to demonstrate differences. In many cases, decisions to purchase one program or another depend on factors other than program qualities.

The Underinvestment in Critical Resources

The changes implied by a goal such as achieving higher levels of scientific literacy for all students require aggressive investment in critical resources. Other factors, such as eroding goals and overuse of innovations, consume capital that could be used to develop physical and human capacity. An example is the lack of investment in professional development of science teachers and the unfocused use of capital for professional development activities that school districts do provide. Educators limit their own ability to improve school science programs through underinvestment in professional development. Although we need new facilities, materials, and equipment, we also need school personnel who make the vital educational connections between those physical aspects of science education and the students.

Developing New Models

Contemplation of achieving scientific literacy presents a number of paradoxes. To have any significant effect, standards would have to be implemented consistently and nationally; yet states and local school districts have the option of not using or only partially using national standards and of creating and implementing their own standards.

National Aspirations and Local Action

When you really consider the aspirations of achieving scientific literacy and being first in the world in science and mathematics by the year 2000, it staggers the mind. Achieving these aspirations is simultaneously admirable and unimaginable. But, then, so was landing a man on the moon. Just as going to the moon was a national goal that required coordination of numerous local actions, so too the goal of achieving scientific literacy requires the combined and coordinated work of science educators in 50 states, 15,000 school systems, and some 100,000 school buildings.

It will not be easy to sustain aspirations of scientific literacy among professionals and the populace. Although we can define scientific literacy, making the case for the importance of developing this goal is illusive. A very important contribution to achieving this goal would be continuous assessment at the local, state, and national levels of our progress toward achieving the goal. The issue of assessment, as I am discussing it, is broader than assessing student achievement. Obviously, we must assess student achievement, and we also must provide an accurate assessment of standards-based systemic reform. The national aspiration of a moon landing was sustained by periodic events, such as sub-orbital manned flight and orbiting the earth, that demonstrated progress toward the goal. It seems that it will be essential to provide intermittent public information on the progress toward our goals, the reasons for progress, and what we must do to sustain our progress. The TIMSS, the NAEP, and the analysis and reporting of state and local assessments should support the continued efforts to achieve higher levels of scientific literacy.

Educational Goals and Political Support

Many statements about achieving scientific literacy are justified in terms that build and maintain political support. For instance, achieving scientific literacy is in the national interest because it will provide a better workforce, continued economic security, and stronger international competitiveness. Such educational aspirations have political support. What seems to vary is the means to activate the goals and the political will to establish one set of large-scale policies, such as national standards, and to stay with those policies for an extended period of time, such as a decade or more.

One insight that emerges from analysis of other large-scale initiatives such as the space race, the war on poverty, and the war on cancer (Schulman, 1980) is the beneficial role of an identifiable "enemy." Perhaps we need an "enemy," a war on ignorance, to galvanize our efforts and sustain public and political support.

Experts and Experience

The scale of reform that we are addressing depends on the expertise of a wide range of individuals. We need scientists, who understand the subtleties of major concepts; researchers, who understand the dynamics of learning; curriculum developers, who understand the complexities of designing instructional materials; assessment specialists, who understand the issues of identifying student achievement; teacher educators, who apply their understanding to the initial knowledge and abilities of classroom teachers; professional developers, who understand the intricacies of adult learning and educational change; and classroom teachers, who understand the issues of teaching and learning in their classrooms. We need all the experts and experience in the professional community working together. What we do not need is the respective individuals and groups assuming they have the expertise in domains other than their own and subsequently asserting their ideas by nature of authority, position, or experience. We need genuine expertise appropriately applied.

We have to overcome the contemporary view that one should be suspicious of any expert and the confounding view that everyone has expertise in all aspects of education. The idea of guiding large-scale educational reform through the coordinated efforts of the professional community is easy to say and difficult to implement because it rests on issues of enlightened professionals, acceptance of expertise, humility in areas where one lacks expertise, and a true spirit of collaboration. Having widespread agreement on what students should know and be able to do is a tremendous first step that can bring unity and coordination to reform efforts.

Continuous Transformation and Single Solution

Improving science education and achieving scientific literacy is, to use Richard Elmore and Milbray McLaughlin's (1988) words, "steady work." For too long those interested in education have applied the "single solution" model: If only we can improve teacher education, change the tests, or get different textbooks, then we will have achieved the reform. The truth is, we need all of these; no single solution will be sufficient in the larger perspective of educational reform. Implementing the model of standards-based, systemic reform also assumes the view of long-term continuous transformation of various components in the educational system. This is a new paradigm for many science educators.

Systemic Reform and Component Projects

Closely related to the problem of perceiving reform in terms of single solutions is identifying component projects, such as a new curriculum, without considering professional development and budgetary support as supporting components for the new curriculum. In our era we have developed a systemic view that requires a much larger perspective and concerted work. A systemic view of reform also implies a more dynamic view of the reform process. Thinking in terms of single solutions and component projects is a more static view that will have to change.

Vertical and Horizontal Integration

In discussions of programs, curricula, and teaching, our first inclination is often to view reform in terms of grade levels, classes, and courses. For example, we think of chemistry for grade 11 or the science curriculum for grade 3. This is what I call horizontal integration. We tend to integrate ideas of content, teaching, and assessment in terms of a grade level, course, or discipline. In some cases, we have established a different, vertical view. For example, the *Scope, Sequence, and Coordination* program of the NSTA incorporates both vertical and horizontal integration for a limited portion of the school science program. The *Middle School Science and*

Technology program of the BSCS also presents both vertical and horizontal perspectives.

There is a need for greater vertical integration in school science programs. There should be common elements, such as inquiry, from kindergarten through grade 12.

The Contemporary Challenges

In the Prologue I asked, Are we sinking, drifting, or sailing? By now the message should be clear. I do not think we are sinking. Some aspects of the science education system may be adrift, but we can and must sail toward the future. In this section, I push the maritime metaphor even further by using a statement from the author and historian Henry Adams. Henry Adams (1958) once said, "The President resembles the commander of a ship at sea, he must have a helm to grasp, a course to steer, a port to seek" (p. 197). Adams could as well have been discussing contemporary educational reform. All of us in the science education community are well served to think about the national standards as a helm to grasp; the alignment of policies, programs, and practices as a course to steer; and the National Education Goals Panel goal number three (steps toward achieving scientific literacy) as a port to seek.

The science education community must shift from an emphasis on formulating standards and achieving a national consensus on what all students should know and be able to do, to more practical and difficult matters of developing and implementing programs and practices aligned with the national standards. The science education community must pause and ask, Where are we in the process of reform? What are the local, state, and federal roles and responsibilities? Where do we go from here? Others have already discussed some of these questions (Ravitch, 1995; Rothman, 1995; Wilson and Davis, 1994).

A Helm to Grasp: National Standards for Science Education

Science teachers, supervisors, curriculum developers, administrators, scientists, and parents can use the vision contained in the *Standards* to improve science education. However, they should recognize that the content standards are not a curriculum, the assessment standards are not examinations, and the teaching standards are not licensing requirements. The details of curriculum, assessment, and licensure are state and local issues and must be addressed in appropriate ways at those levels. The *Standards* do present a vision of changes and improvements for science education and goals for student achievement. Science educators in professional organizations, 50 state departments of education, 15,000 school districts, and close to two million teachers of science will have to translate the standards into curriculum, assessment, and teaching programs and practices.

The science education community also will benefit from the fact that the

Standards provides support for the integrity of school science programs. The *Standards* presents science as a way of knowing that is based on such things as empirical criteria, logical argument, and skeptical review. Although individuals have ways of knowing that are subjective, qualitative, and personal, these views are rarely supported with evidence and logic and are usually not open to question. Such views may be personally meaningful, but they are not scientific. In recent years, various groups have presented views that they claim are scientific but on analysis are not. In fact, some are antiscientific views (Gross and Levitt, 1994; Holton, 1993). The *Standards* supports the science teachers, supervisors, administrators, and communities who understand that such positions have no place in school science programs.

National standards for science education represent a helm to grasp. But, the development of national standards completes only the first step in the course of standards-based systemic reform of science education. Contemporary educational thought (Fullan and Stiegelbauer, 1991; Hall, 1992; Fullan and Miles, 1992; Hall and Loucks, 1978; Loucks-Horsley and Stiegelbauer, 1991) supports the conclusion that the science education community needs more than standards to initiate and sustain the changes outlined in the *Standards*. There are several reasons for this conclusion. First, although the *Standards* is well thought out, provides accurate descriptions of content, and includes some examples of curriculum instruction and assessment, it does not lend itself to easy interpretation. Second, the *Standards* does not provide descriptions of processes for translating the *Standards*. It is not, for example, a manual on the development of curriculum materials. Describing the process of implementing the *Standards* was not the purpose of the NRC project. The observation, however, does suggest the need for discussions and illustrations for those educators who must use the *Standards*. Third, the changes implied by the national standards present a complex array of interdependent factors involving content, teaching, assessment, professional development, school science programs, and systemic reform. Examining and understanding a document such as the *Standards* as an integrated set of policies is neither intuitively obvious nor commonly done in the experience of most educators. Finally, educators are bound by their current views of science education programs and practices. These views often are contrary to the spirit of the *Standards*. As a result, some educators claim that current practices already align with the *Standards,* and some commercial publishers complete a cursory review of the *Standards* and a cosmetic revision of materials and claim their program is standards-based.

These factors imply the need for professional dialogue and exemplary illustrations of what the *Standards* means in terms of specific components of science education, for example, curriculum materials and assessments, and what they imply in general for school science programs, professional development, and systemic reform. There is a need to provide expressions of the *Standards* through exemplars that bring the *Standards* to life so that it can guide those responsible for translating and implementing standards.

From the perspective of school science programs, the *Standards* recommends but does not provide coherent and coordinated instructional sequences and associated assessments that constitute opportunities to learn the content and evidence of student achievement. This is not a weakness of the *Standards;* rather, it is a strength. The *Standards* represents a large-scale policy initiative that specifically avoids the unintentional consequences of framing a national curriculum.

The science education community has a task of grasping the helm and beginning to steer the course of science education. In brief, we must unify our support of standards. Using the national standards will provide a much needed stability, coordination, and consistency to reform. At the same time, we must realize that grasping the helm does not assure a safe and successful voyage. We must look beyond and begin navigating a course that includes development of science programs and implementation of classroom practices, both of which must be supported through assessments, professional development, and systemwide alignment of resources.

A Course to Steer: Aligning Purposes, Policies, Programs, and Practices

Concerning a course to steer and an assessment for our location on the journey, I return to the framework that forms one theme of this book. This framework describes different dimensions of reform efforts, whether that reform is at the local, state, or national level. In the briefest form, the dimensions of reform involve clarifying the *purposes* of science education, establishing *policies* for different aspects of science education, developing *programs* for science education, and changing *practices* in science classrooms (Bybee, 1993, 1995, 1996).

The term *purposes* includes aims, goals, and a rationale for science education. Achieving scientific literacy, for example, is an example of a purpose. At the national level, the publication *Science for All Americans* (Rutherford and Ahlgren, 1989) provides details and categories of scientific literacy.

Policy statements are concrete translations of the purpose and apply to subsystems, such as disciplines, teacher education, and grade levels within local school districts. At the national level, the *National Science Education Standards* (National Research Council, 1996a) serves as an example of a policy document. Although many science educators criticize policies, these do have an essential role in any reform because they provide links between purpose and programs, and those links help bring about greater consistency and coherence with the science education system.

Programs are the actual materials, textbooks, assessments, and equipment that are based on policies and developed to implement the stated policies. Programs are unique to different aspects of science education, such as grade levels and disciplines. For example, the BSCS curricula *Science for Life and Living, Middle School Science and Technology,* and *Biological Science: A Human Approach* for elementary, middle, and high school, respectively, are examples of this dimension of reform. This example serves to point out the strength of policies. If you compare the NSRC program *Science and Technology for Children;* the NSTA program *Scope, Sequence, and Coordination;* and the EDC, *Insights in Biology,* you can see that there are differences between these programs and the BSCS programs. This contrast indicates the variations that can occur in the translation from a general set of policies to specific programs. Assessments, such as NAEP, Iowa Test of Basic Skills,

American College Testing, the California Achievement Test, Scholastic Aptitude Test, and other state and locally developed assessments also fall in the program category.

Practice focuses on the specific actions of science educators. Practice is the most unique and fundamental dimension, and it is based on the educator's understanding of the purpose, objectives, curriculum, school, learners, and his or her strengths as a teacher. Examples of this dimension are unique to teachers' actions and behaviors in the classroom.

If one uses the framework of purpose, policy, program, and practice, and does an analysis of issues such as time for change, number of individuals involved in changing a particular dimension, the scope of change, duration of change once it has occurred, and the difficulty of reaching agreement on the changes, it becomes evident that our current situation is at the interface between policy and program, and we are steering a course toward greater challenges. The major trend is toward the improvement of science education programs and practices. The issues associated with this trend include the alignment of curriculum materials, assessments, practices, and professional development. These issues of alignment must receive support at school, state, and national levels for the changes implied by standards.

The perspective in this section complements the systemic view of contemporary reform. In many cases, systemic initiatives have identified the various components within the science education system. Work is beginning on improvements through various policy, programmatic, and practical initiatives. The framework of purpose, policy, program, and practice provides a course to steer for the various components.

The education system has individuals, agencies, and organizations that work on virtually all dimensions of the purpose to practice framework. A critical issue in this reform centers on 4 Cs—coordination, consistency, coherence, and collaboration—among various components in the science education system.

A Port to Seek: Higher Levels of Achieving Scientific Literacy for All Students

> By the year 2000, American students will leave grades four, eight, and twelve having demonstrated competency in challenging subject matter including . . . science[2]; and every school in America will ensure that all students learn to use their minds well, so they may be prepared for responsible citizenship, further learning, and productive employment in our modern economy.

This goal is from the National Education Goals formulated at the historic summit on education held in Charlottesville, Virginia. At the summit, the president and governors declared, "The time has come, for the first time in United States history, to establish clear national performance goals, goals that will make us internationally competitive." Note that my emphasis is on goal number three, student achievement

---

[2]I have only eliminated reference to other disciplines.

and citizenship, not on goal number four, U.S. students will be first in the world in science and mathematics achievement. Although the latter makes for good political slogans and a succinct statement, almost anyone who has thought about the goal and tried to translate it to programs and practices has confronted a labyrinth of problems that seem insurmountable. I propose that goal number three presents a more reasonable, focused, and achievable goal. Even with this goal, the challenges are formidable, but I would argue they are feasible. They also are remarkably consistent with the *National Science Education Standards.*

Following are several reasons it may be important to focus on goal number three. First, the grade levels 4, 8, and 12 align with the Science Content Standards, and this provides science teachers, curriculum developers, science supervisors, assessment specialists, and professional developers clear and appropriate content for elementary, middle, and high school programs. Second, the issue of demonstrating competence will be up to national, state, and local prerogatives, not other nations and international teams of assessment specialists. Third, the aim of ensuring that all students learn to use their minds well directly relates to the emphasis on inquiry and design in the *Standards.* That is, using inquiry and design as stated in the *Standards* should enhance all students' cognitive development. Finally, broadening the scope of content to include technology, science in personal and social perspectives, and history and nature of science will prepare students for responsible citizenship, further learning, and productive employment. My recommendation to shift our focus may not be politically correct, but I do think it may be practically appropriate.

In many respects, the stage is set to move beyond standards and benchmarks. We are reporting on annual progress toward Goals 2000 (National Education Goals Panel, 1994); we are revising state frameworks (Council of Chief State School Officers, 1995); we are initiating systemic initiatives (SRI International 1995); we are providing regional support and technical assistance to improve science and mathematics education (SRI International 1995).

Relative to implementing the *Standards,* all individuals in the science education community have a helm to grasp and a course to steer, and the *Standards* should become an emphasis as science teachers concentrate on learners in their classrooms; school districts provide support and resources for their teachers; states provide appropriate curriculum frameworks, assessment programs, and opportunities for professional development; and federal agencies implement policies that will provide support to align curriculum, assessment, and teaching practices with the *Standards.*

Conclusion

Elsewhere I have suggested that reform is best served from a perspective of distributed leadership, not of a few key people making all the decisions (Bybee, 1993). Such a view implies that all individuals in the science education community have a role in reform. It is too easy to place responsibility and blame elsewhere and wait for reform to happen. Hope lies in each individual's assuming that he or she is a key decision maker for some aspect of reform.

Obstacles to reform include contemporary politics and economics and the failure of the public to recognize and support the critical role of education in our society. Also, educators clearly have done very little to enhance the public's understanding of education. Educators have embraced every slogan, trend, and quick-fix imaginable, only to find that they all fail to solve the problems.

Effective leadership in this reform must meet several indispensable recommendations. The first requirement is to point the entire nation in one direction. Beginning with the 1990 Governors Conference and continuing with President Clinton's America 2000, we have had leadership in the White House and subsequently in the U.S. Department of Education, NSF, NRC, AAAs, and public and private agencies that have continued to steer the nation in one direction. Most immediately, this requirement suggests the need to support the *Standards* and use them as central to the reform of curriculum, assessment, and instruction for schools and classroom teachers.

The 1996 Conference of Governors and Business Leaders reaffirmed the goal and role of standards. It left open the 4 Cs issue of consistency, coordination, coherence, and collaboration. The education community will either discover the national standards or reinvent them, because the standards establish a critical conflict between this nation and our global competition.

A second recommendation involves explaining to citizens why the direction proposed in the *Standards* is right for the nation. Here the obligation falls to public officials and especially to scientists, science educators, and science teachers involved in science education at local, state, and national levels. For many people, it is not obvious why standards will improve the science education of their children. Indeed, there have been some criticisms and misconceptions of what standards are and what standards can do.

A third recommendation entails support for those who must implement the changes implied in the *Standards*. The reform will be judged a failure if we do not develop a system that continually supports the various initiatives that must occur at all levels in the educational system. A requirement of support for the reform assumes that standards will be a central feature and that programs, as I have discussed them, will be available for science teachers.

In moving beyond standards and sustaining the reform of science education, we encounter the challenges of developing science education programs and improving teaching practices. Most individuals who realize the importance of these changes also recognize the time and difficulties involved. Although the year 2000 is rapidly approaching, the Achilles heel of this reform clearly lies in the expectation that the reform will be easy and quick. This said, the *Standards* gives us a helm to hold; the framework of purpose, policies, programs, and practices presents a course to steer; and assuming individual responsibility for our part of achieving scientific literacy for all students provides a port to seek.

EPILOGUE

For twenty decades, teachers have continually endeavored to achieve the goal of scientific literacy; they have educated students in the sciences and contributed to the educational purposes of a developing nation that needs both scientists and enlightened laypeople. Almost four decades ago, we entered an international race to send men to the moon and return them home safely. That was the "Golden Age" of curriculum reform in science education. That reform is now past, but the goal of achieving scientific literacy remains. Our journey toward scientific literacy has been long. And the destination has varied as different national needs have emerged.

A decade ago, we entered an era of educational reports. Every group with the power to meet and the ability to prepare a document made recommendations. Collectively, the nation called 911 and reported a crisis in education.

In response, the educational community initiated reform. Science educators must provide programs for all students—those of high aptitude and low, girls and boys, minority and majority, physically able and disabled. The invitation to science education is clear; now the response must come from all of us. Lately, we have seen problems in our economy deepen, concerns about our environment widen, democratic freedoms challenged, and political winds change directions. Our era is new and the challenges are significant—maintaining a nation that is economically productive, ecologically sustainable, and globally harmonious. Science education can meet these challenges by developing citizens who understand basic scientific and technological principles, who understand the role and history of science and technology in society, who value the contributions and limitations of science and technology, and who take personal and social actions based on the principles and processes of science and technology.

Our present programs and practices fail the test of scientific, educational, and social adequacy. Just as historical leaders like Jerrold Zacharias, Robert Karplus, Joseph Schwab, Marjorie Gardner, William Mayer, J. Darrell Barnard, James B. Conant, Gerald Craig, Bertha Parker, Anna Botsford Comstock, Charles Eliot,

Frances W. Parker, and Mary Budd Rowe took up historical challenges, now we are summoned to lead a new reform of science education.

In less than a decade, we will be called to account for the achievements outlined in reform documents such as national standards. One or two or three leaders, even if they are great, cannot and will not accomplish the task of reforming science education. One or two or three new policy statements, science programs, or assessment packages will not successfully improve school science. The scale of the system is too large, the dispersion of power too wide, and the variation of students, teachers, and schools too great.

If we are to succeed, the leadership for this reform must come from the whole science education community. We must internalize and act on the message—each of us is the reform. Each individual must assume responsibility for improving science education through educating science teachers, developing curriculum materials, supervising state and local reforms, researching teaching practices, and most important, teaching science to the boys and girls, early adolescents, and young adults who will be scientists and informed laypeople.

Just as our era is new, so our models of reform, our view of achieving scientific literacy, and our role in accommodating the challenges also must be new. Historical approaches of single-topic workshops, curriculum materials, tests, summer institutes, make-it and take-it, and textbook adoptions must evolve into teaching and learning, professional development, appropriate assessment strategies, standards-based programs, and systemic approaches to educational improvement.

In future decades, history will not well remember the hundreds of policy statements that rallied us to this reform. History will record our accomplishments. The first chapter of that history is being written now, and it will be published in the year 2000. We must set our sights on the future and design programs and implement practices that help us sail toward the purpose of achieving scientific literacy.

REFERENCES

Abraham, M. R., and J. W. Renner. 1986. "The Sequence of Learning Cycle Activities in High School Chemistry." *Journal of Research in Science Teaching* 23: 121–143.

Adams, H. 1958. "The Session, 1869–1870." In *The Great Secession Winter of 1860–61 and Other Essays,* ed. G. E. Hochfield. New York.

Adler, M. J. 1981. *Six Great Ideas.* New York: Macmillan.

———. 1982. *The Paideia Proposal: An Educational Manifesto.* New York: Macmillan.

———. 1984. *A Vision of the Future.* New York: Macmillan.

———. 1987. *We Hold These Truths.* New York: Macmillan.

Agin, M. 1974. "Education for Scientific Literacy: A Conceptual Frame of Reference and Some Applications." *Science Education* 58: 3.

Agin, M. L., and M. O. Pella. 1972. "Teaching Interrelationships of Science and Society Using a Socio-Historical Approach." *School Science and Mathematics* 72: 320–334.

Aldridge, W. G. 1989. *Essential Changes in Secondary School Science: Scope, Sequence, and Coordination.* Washington, DC: National Science Teachers Association.

American Association for the Advancement of Science. 1988. *Science, Engineering, and Ethics. A Report on an American Association for the Advancement of Science Workshop and Symposium.* Washington, DC: American Association for the Advancement of Science.

———. 1993. *Benchmarks for Science Literacy.* Washington, DC: American Association for the Advancement of Science.

American Psychological Association and Mid-Continent Regional Educational Laboratory. 1993. *Learner-Centered Psychological Principles: Guidelines for School Redesign and Reform.* Denver: Mid-Continent Regional Educational Laboratory.

Anderson, A. 1993. "The Stages of Systemic Change." *Educational Leadership* 51(1): 14–17.

Anderson, R. 1983. "Are Yesterday's Goals Adequate for Tomorrow?" *Science Education* 67(2): 171–176.

————. 1994. *Issues of Curriculum Reform in Science, Mathematics and Higher-Order Thinking Across the Disciplines.* Washington, DC: U.S. Department of Education, Office of Educational Research and Improvement.

Arons, A. B. 1983. "Achieving Wider Scientific Literacy." *Daedalus: Journal of the American Academy of Arts and Sciences* 112(2): 91–122.

Atkin, J. M., and R. Karplus. 1962. "Discovery or Invention." *The Science Teacher* 29(2): 121–143.

Bady, R. 1979. "Students' Understanding of the Logic of Hypothesis Testing." *Journal of Research in Science Teaching* 16: 61–65.

Banathy, B. 1991. *Systems Design of Education.* Englewood Cliffs, NJ: Educational Technology Publications.

————. 1992. *A Systems View of Education: Concepts and Principles for Effective Practice.* Englewood Cliffs, NJ: Educational Technology Publications.

Barnard, J. D. 1960. *Rethinking Science Education.* Chicago: National Society for the Study of Education, University of Chicago Press.

Barney, G. 1980. *The Global 2000 Report to the President: Entering the Twenty-first Century.* Washington, DC: U.S. Government Printing Office.

Barzun, J. 1991. *Begin Here: The Forgotten Condition of Teaching and Learning.* Chicago: University of Chicago Press.

Berkheimer, G., and C. W. Anderson. 1989. *The Matter and Molecules Project: Curriculum Development Based on Conceptual Change Research.* Paper presented at the Annual Meeting of the National Association for Research in Science Teaching, San Francisco, CA.

Berliner, D. C., and B. J. Biddle. 1995. *The Manufactured Crisis: Myths, Fraud, and the Attack on America's Public Schools.* Reading, MA: Addison-Wesley.

Bertalanfy, L. von. 1968. *General Systems Theory.* New York: George Braziller.

Billeh, V., and O. Hassan. 1975. "Factors Affecting Teachers' Gain in Understanding the Nature of Science." *Journal of Research in Science Teaching* 12: 209–219.

Biological Sciences Curriculum Study. 1989. *New Designs for Elementary School Science and Health.* Dubuque, IA: Kendall/Hunt.

————. 1993. *Developing Biological Literacy: A Guide to Developing Secondary and Post-Secondary Biology Education.* Dubuque, IA: Kendall/Hunt.

Black, P., and J. M. Atkin, eds. 1996. *Changing the Subject: Innovations in Science, Mathematics and Technology Education.* New York: Organization for Economic Co-operation and Development.

Bloom, A. 1987. *The Closing of the American Mind.* New York: Simon and Schuster.

Bowyer, J., and M. Linn. 1978. "Effectiveness of the Science Curriculum Improvement Study in Teaching Scientific Literacy." *Journal of Research in Science Teaching* 15(3): 209–219.

Boyer, E. L. 1983. *High School: A Report on Secondary Education in America.* New York: Harper and Row.

Boyer, E. L., and A. Levine. 1981. *A Quest for Common Learning: The Aims of General Education.* Princeton, NJ: Carnegie Foundation for the Advancement of Teaching.

Bracey, G. W. 1996. "International Comparisons and the Condition of American Education." *Educational Researcher* 25(1): 5–11.

Bragaw, D., and M. Hartoonian. 1988. "Social Studies: The Study of People in Society." In *The Content of the Curriculum,* ed. R. S. Brandt. Alexandria, VA: Association for Supervision and Curriculum Development.

Bransford, J. D., and N. J. Vye. 1989. "A Perspective on Cognitive Research and Its Implications for Instruction." In *Toward the Thinking Curriculum: Current Cognitive Research, 1989 Yearbook of the ASCD,* ed. L. Resnick and L. Klopfer. Alexandria, VA: Association for Supervision and Curriculum Development.

Brown, F. 1973. *The Reform of Secondary Education.* New York: McGraw-Hill.

Brown, L., A. Durning, C. Flavin, H. French, J. Jacobson, N. Lenssen, M. Lowe, S. Postel, M. Renner, L. Starke, P. Weber, and J. Young. 1993. *State of the World: 1993.* New York: W. W. Norton.

Brown, L., A. Durning, C. Flavin, J. Jacobson, N. Lenssen, M. Lowe, S. Postel, M. Renner, L. Starke, P. Weber, and J. Young. 1984. *State of the World: 1984.* New York: W. W. Norton.

———. 1985. *State of the World: 1985.* New York: W. W. Norton.

———. 1986. *State of the World: 1986.* New York: W. W. Norton.

———. 1987. *State of the World: 1987.* New York: W. W. Norton.

———. 1988. *State of the World: 1988.* New York: W. W. Norton.

———. 1989. *State of the World: 1989.* New York: W. W. Norton.

———. 1990. *State of the World: 1990.* New York: W. W. Norton.

———. 1991. *State of the World: 1991.* New York: W. W. Norton.

———. 1992. *State of the World: 1992.* New York: W. W. Norton.

Brown, S. 1977. "A Review of the Meanings of and Arguments for Integrated Science." *Studies in Science Education* 4:31–62.

Bruer, J. T. 1993. *Schools for Thought: A Science of Learning in the Classroom.* Cambridge, MA: MIT Press.

Bruner, J. 1960. *The Process of Education.* New York: Vintage.

Brush, S. 1984. *History of Modern Science: Teachers Guide.* College Park, MD: University of Maryland.

Burke, J. 1978. *Connections.* Boston: Little, Brown.

Bush, G. 1991. *America 2000: An Educational Strategy.* Washington, DC: U.S. Office of Education.

Bush, V. 1945. *Science: The Endless Frontier.* Washington, DC: U.S. Government Printing Office. (Reprinted in 1980, New York: Arno Press)

Bybee, R. W. 1977. "The New Transformation of Science Education." *Science Education* 61(1): 85–97.

———. 1979. "Science Education and the Emerging Ecological Society." *Science Education* 63: 95–109.

———. 1982. "Citizenship and Science Education." *The American Biology Teacher* 41(3): 154–163.

———. 1985. "The Restoration of Confidence in Science and Technology Education." *School Science and Mathematics* 85(2): 95–108.

———. 1986. "The Sisyphean Question in Science Education: What Should the Scientifically and Technologically Literate Person Know, Value, and Do—As a Citizen?" In *Science-Technology-Society 1985 NSTA Yearbook,* ed. R. W. Bybee, 79–93. Washington, DC: National Science Teachers Association.

———. 1987. "Science Education and the Science-Technology-Society (STS) Theme." *Science Education* 71(5): 668–683.

———. 1989. "Teaching High School Biology: Materials and Strategies." In *High School Biology: Today and Tomorrow,* ed. W. G. Rosen, 165–177. Washington, DC: National Academy Press.

———. 1993. *Reforming Science Education: Social Perspectives and Personal Reflections.* New York, NY: Teachers College Press.

———. 1994. *Reforming Science Education: Personal and Social Perspectives.* New York: Teachers College Press.

———. 1995. "Achieving Scientific Literacy." *The Science Teacher,* 62(7): 28–33.

———. 1996. "The Contemporary Reform of Science Education." In *Issues in Science Education,* ed. J. Rhoton and P. Bowers, 1–14. Arlington, VA: National Science Teachers Association.

———. 1996. "Improving Science Education." In *National Standards & the Science Curriculum: Challenges, Opportunities, & Recommendations,* ed. R. W. Bybee, 41–54. Dubuque, IA: Kendall/Hunt Publishing.

———. 1996. "Science Education Beyond Standards: Improving Programs and Practices." In *National Standards & the Science Curriculum: Challenges, Opportunities, & Recommendations,* ed. R. W. Bybee, 53. Dubuque, IA: Kendall/Hunt Publishing.

Bybee, R. W., C. E. Buchwald, S. Crissman, D. R. Heil, P. J. Kuerbis, C. Matsumoto, and J. D. McInerney. 1989. *Science and Technology Education for the Elementary Years: Frameworks for Curriculum and Instruction.* Washington, DC: National Center for Improving Science Education.

———. 1990. *Science and Technology Education for the Middle Years: Frameworks for Curriculum and Instruction.* Washington, DC: The National Center for Improving Science Education.

Bybee, R. W., A. B. Champagne, S. Loucks-Horsley, S. A. Raizen, and P. J. Kuerbis. 1990a. *Getting Started in Science: A Blueprint for Science Education in the Middle Years.* Washington, DC: The National Center for Improving Science Education.

———. 1990b. *Assessment in Science Education: The Middle Years.* Washington, DC: The National Center for Improving Science Education.

Bybee, R. W., and G. DeBoer. 1993. "Goals for the Science Curriculum." In *Handbook of Research on Science Teaching and Learning.* Washington, DC: National Science Teachers Association.

Bybee, R. W., J. D. Ellis, J. R. Giese, and L. S. Parisi. 1992a. *Teaching About the History and Nature of Science and Technology: A Curriculum Framework.* Colorado Springs, CO: Biological Sciences Curriculum Study.

———. 1992b. *Teaching About the History and Nature of Science and Technology: Background Papers.* Colorado Springs, CO: Biological Sciences Curriculum Study.

Bybee, R. W., J. D. Ellis, and M. R. Matthews, eds. 1992. "Teaching About the History and Nature of Science and Technology: An Introduction." *Journal of Research in Science Teaching* 29(4): 327–329.

Bybee, R. W., P. D. Hurd, J. B. Kahle, and R. Yager. 1981. "Human Ecology: An Approach to the Science Laboratory." *The American Biology Teacher* 43(6): 304–311; 326.

Bybee, R. W., and P. L. Legro. 1996. Developing and Implementing Standards to Improve Science Education: Perspective on the U.S. Experience. A Concept Paper prepared for The World Bank Human Development Department, Washington, DC.

Bybee, R. W., and J. D. McInerney, eds. 1995. *Redesigning the Science Curriculum: A Report on the Implications of Standards and Benchmarks for Science Education.* Colorado Springs, CO: Biological Sciences Curriculum Study.

Bybee, R. W., J. C. Powell, J. D. Ellis, J. R. Giese, L. S. Parisi, and L. Singleton. 1991. "Integrating the History and Nature of Science and Technology in Science and Social Studies Curriculum." *Science Education* 75(1): 143–155.

Bybee, R. W., and R. B. Sund. 1982. *Piaget for Educators.* Columbus, OH: Charles E. Merrill Publishing Company.

California Department of Education. 1990. *Science Framework for California Public Schools.* Sacramento, CA: State Department of Education.

Campbell, D. T., and J. C. Stanley. 1963. *Experimental and Quasi-experimental Designs for Research.* Boston: Houghton Mifflin.

Cannon, J., and J. Jinks. 1992. "A Cultural Literacy Approach to Assessing General Scientific Literacy." *School Science and Mathematics* 7(4): 96–200.

Carey, R. L., and N. Strauss. 1968. "An Analysis of the Understanding of the Nature of Science by Prospective Secondary Science Teachers." *School Science and Mathematics* 70: 366–376.

Carlton, R. 1963. "On Scientific Literacy." *NEA Journal* 52: 33–35.

Carson, R. 1962. *Silent Spring.* Boston: Houghton Mifflin.

Caswell, H., and D. Campbell. 1935. *Curriculum Development.* New York: American Book.

Champagne, A. B. 1987. *The Psychological Basis for a Model of Science Instruction.* A commissioned paper for IBM-supported Design Project. Colorado Springs, CO: Biological Sciences Curriculum Study.

Champagne, A. B., ed. 1988. *This Year in School Science 1988: Science Teaching, Making the System Work.* Washington, DC: American Association for the Advancement of Science.

Champagne, A. B., and L. E. Hornig, eds. 1985. *This Year in School Science 1985: Science Teaching.* Washington, DC: American Association for the Advancement of Science.

———. 1986. *This Year in School Science 1986: The Science Curriculum.* Washington, DC: American Association for the Advancement of Science.

———. 1987. *This Year in School Science 1987: Students and Science Learning.* Washington, DC: American Association for the Advancement of Science.

Champagne, A. B., and B. E. Lovitts. 1989. "Scientific Literacy: A Concept in Search of a Definition." In *This Year in School Science 1989: Scientific Literacy,* ed. A. B. Champagne and B. E. Lovitts. Washington, DC: American Association for the Advancement of Science.

Champagne, A. B., and B. E. Lovitts, eds. 1989. *This Year in School Science 1989: Scientific Literacy.* Washington, DC: American Association for the Advancement of Science.

Champagne, A. B., S. Loucks-Horsley, P. J. Kuerbis, and S. A. Raizen. 1991. *Science and Technology Education for the High School Years.* Washington, DC: National Center for Improving Science Education.

Champagne, A. B., and S. A. Raizen. 1991. *The High Stakes of High School Science.* Washington, DC: National Center for Improving Science Education.

Chen, D., and P. Novik. 1986. *Scientific and Technological Literacy for All: An Israeli Educational Challenge for the Future.* Tel Aviv: Science Education Center, Tel Aviv University.

Cheney, L. V. 1987. *American Report: A Report on the Humanities in the Nation's Public Schools.* Washington, DC: Superintendent of Documents.

Churchman, C. W. 1979. *The Systems Approach.* New York: Delta.

Clement, J. 1987. "Overcoming Students' Misconceptions in Physics: The Role of Anchoring Intuitions and Analogical Validity." In *Proceedings of Second International Seminar: Misconceptions and Educational Strategies in Science and Mathematics III,* ed. J. Novak. Ithaca, NY: Cornell University.

Clewell, B., B. Anderson, and M. Thorpe. 1992. *Breaking the Barriers: Helping Female and Minority Students.* San Francisco: Jossey-Bass.

Cohen, D. K. 1995. "What Is the System in Systemic Reform?" *Educational Researcher* 24(9): 11–17, 31.

Cohen, D. K., and J. P. Spillane. 1993. "Policy and Practice: The Relations Between Governance and Instruction." In *Designing Coherent Education Policy: Improving the System,* ed. S. H. Fuhrman, 35–95. San Francisco: Jossey-Bass.

Cohen, I. B., and F. G. Watson, eds. 1952. *General Education in Science.* Cambridge, MA: Harvard University Press.

Cole, M., and P. Griffin. 1987. *Contextual Factors in Education: Improving Science and Mathematics Education for Minorities and Women.* Madison, WI: Wisconsin Center for Education Research.

Coleman, J. 1966. *Equality of Educational Opportunity.* Washington, DC: U.S. Government Printing Office.

———. 1974. *Youth: Transition to Adulthood.* Chicago: University of Chicago Press.

Collins, A. 1989. "Elementary School Science Curricula That Have Potential to Promote Science Literacy (And How to Recognize One When You See One)." In *The Year in School Science 1989: Scientific Literacy,* ed. A. B. Champagne and B. E. Lovitts. Washington, DC: American Association for the Advancement of Science.

———. 1995. "National Science Education Standards in the United States: A Process and a Product." *Studies in Science Education* 26: 7–37.

Commission on the Humanities. 1980. *The Humanities in American Life.* Berkeley, CA: University of California Press.

Commission on the Reorganization of Secondary Education. 1918. *The Cardinal Principles of Secondary Education.* Washington, DC: U.S. Bureau of Education.

Commission on Secondary School Curriculum. 1937. *Science in General Education.* New York: Appleton-Century-Crofts.

Committee on Curriculum Studies. 1971. "School Science Education for the 70s." *The Science Teacher* 38: 46–51.

Committee on the Function of Science in General Education. 1938. *Science in General Education.* New York: Appleton-Century-Crofts.

Committee on History and Social Studies. 1988. *History-Social Science Framework for California Public Schools: Kindergarten through Grade Twelve.* Sacramento, CA: State Department of Education.

Compayre, G. 1907. *Herbart and Education by Instruction,* trans. M. Findlay. New York: Crowell.

Conant, J. 1946. "The Science Education of the Layman." *Yale Review* 36: 15–36.

———. 1959. *The American High School Today.* New York: McGraw-Hill.

———. 1961. *Slums and Suburbs.* New York: McGraw-Hill.

———. 1963. *The Education of American Teachers.* New York: McGraw-Hill.

———. 1964. *Shaping Educational Policy.* New York: McGraw-Hill.

Connelly, F. M. 1969. "Philosophy of Science and Science Curriculum." *Journal of Research in Science Teaching* 8: 108–113.

Cooley, W., and L. Klopfer. 1963. "The Evaluation of Specific Educational Innovations." *Journal of Research in Science Teaching* 8: 73–80.

Corn, J. J. 1966. *Imagining Tomorrow: History, Technology, and the American Future.* Cambridge, MA: MIT Press.

Costenson, K., and A. Lawson. 1986. "Why Isn't Inquiry Used in More Classrooms?" *The American Biology Teacher* 48(3): 150–158.

Council of Chief State School Officers. 1995a. *State Curriculum Frameworks in Mathematics and Science: Results from a Fifty-State Study (Review Draft).* Washington, DC: Council of Chief State School Officers.

———. 1995b. *State Curriculum Frameworks in Mathematics and Science: How Are They Changing Across the States?* Washington, DC: Council of Chief State School Officers.

Cremin, L. 1965. *The Genius of American Education.* New York: Vintage.

———. 1975. "Public Education and the Education of the Public." *Teachers College Record* 77(1): 1–12.

———. 1988. *American Education: The Metropolitan Experience.* New York: Harper and Row.

———. 1990. *Popular Education and Its Discontents.* New York: Harper and Row.

Daugs, D. 1970. "Scientific Literacy—Reexamined." *The Science Teacher* 37(8): 10–11.

DeBoer, G. E. 1991. *A History of Ideas in Science Education: Implications for Practice.* New York: Teachers College Press, Columbia University.

DeBoer, G. E., and R. W. Bybee. 1995. "The Goals of Science Curriculum." In *Redesigning the Science Curriculum: A Report on the Implications of Standards and Benchmarks for Science Education,* ed. R. W. Bybee and J. D. McInerney, 71–74. Colorado Springs, CO: Biological Sciences Curriculum Study.

deCastell S., and A. Luke. 1986. "Models of Literacy in North American Schools: Social and Historical Conditions and Consequences." In *Literacy, Society, and Schooling,* ed. S. deCastell, A. Luke, and K. Egan. Cambridge: Cambridge University Press.

deCastell, S., A. Luke, and K. Egan. 1986. *Literacy, Society, and Schooling.* Cambridge: Cambridge University Press.

deCastell S., A. Luke, and D. MacLennar. 1986. "On Defining Literacy." In *Literacy, Society, and Schooling,* ed. S. deCastell, A. Luke, and K. Egan. Cambridge: Cambridge University Press.

Dede, C., and J. Hardin. 1973. "Reforms, Revisions, Reexaminations: Secondary Science Education Since World War II." *Science Education* 57(4): 485–491.

Dewey, J. 1916. *Democracy and Education.* New York: Macmillan.

———. 1933 [1910]. *How We Think.* Lexington, MA: D. C. Heath.

di Sessa, A. 1982. "Unlearning Aristotelian Physics: A Study of Knowledge-Based Learning." *Cognitive Science* 6: 37–75.

Dow, P. 1991. *Schoolhouse Politics.* Cambridge, MA: Harvard University Press.

Driver, R. 1989. "Students Conceptions and the Learning of Science: Introduction." *International Journal of Science Education* 11(5): 481–490.

Driver, R., E. Guesne, and A. Tiberghien, eds. 1985. *Children's Ideas in Science.* Philadelphia: Open University Press.

Driver R., and V. Oldham. 1986. "A Constructivist Approach to Curriculum Development in Science." *Studies in Science Education* 13:105–122.

Driver, R., A. Squires, P. Rushworth, and V. Wood-Robinson. 1994. *Making Sense of Secondary Science: Research into Children's Ideas.* London: Routledge.

Duschl, R. 1985. "Science Education and Philosophy of Science: Twenty-five Years of Mutually Exclusive Development." *School Science and Mathematics* 85: 541.

———. 1988. "Abandoning the Scientific Legacy in Science Education." *Science Education* 72: 51–62.

———. 1990. *Restructuring Science Education.* New York: Teachers College Press, Columbia University.

Duschl, R., and R. Hamilton. 1992. *Philosophy of Science Cognitive Psychology and Educational Theory and Practice.* Albany: State University of New York Press.

Eckholm, E. 1982. *Down to Earth: Environment and Human Needs.* New York: W. W. Norton.

Educational Policies Commission. 1944. *Education for All American Youth.* Washington, DC: National Education Association.

Egan, D. E., and J. G. Greeno. 1973. "Piagetian Theory and Instruction in Physics." *The Physics Teacher* 11(3):165–169.

Ehrlich, P. R., C. Sagan, D. Kennedy, and W. O. Roberts. 1984. *The Cold and the Dark: The World After Nuclear War.* New York: W. W. Norton.

Eisenhart, M., E. Finkel, and S. Marion. 1996. "Creating the Conditions for Scientific Literacy: A Reexamination." *American Educational Research Journal* 33(2): 261–295.

Eisenkraft, A. 1995. "Equity and Excellence." In *Redesigning the Science Curriculum: A Report on the Implications of Standards and Benchmarks for Science Education,* ed. R. W. Bybee and J. D. McInerney, 81–83. Colorado Springs, CO: Biological Sciences Curriculum Study.

Eisner, E. W. 1975. *The Perceptive Eye: Toward the Reformation of Educational Evaluation.* Stanford, CA: Stanford Evaluation Consortium.

Elmore, R. F. 1996. "Getting to Scale with Good Educational Practice." *Harvard Educational Review* 6(1): 1–26.

Elmore, R. F., and S. H. Fuhrman. 1994. *The Governance of Curriculum, 1994 Yearbook of the ASCD*. Alexandria, VA: Association for Supervision and Curriculum Development.

Elmore, R., and M. McLaughlin. 1988. *Steady Work: Policy, Practice, and the Reform of American Education*. Santa Monica, CA: RAND Corporation.

Engineering Concepts Curriculum Project. 1971. *The Man-Made World*. New York: McGraw-Hill.

Evans, H. 1962. "Towards Scientific Literacy." *Teachers College Record* 64(1): 74–79.

Evans, T. 1970. "Scientific Literacy: Whose Responsibility?" *The American Biology Teacher* 32(8): 80–84.

Faunce, W. A. 1981. *Problems of an Industrial Society*. New York: McGraw-Hill.

Fensham, P. J. 1985. "Science for All: A Reflective Essay." *Journal of Curriculum Studies* 17(4): 415–435.

———. 1986/87. "Science for All." *Educational Leadership* 44: 18–23.

———. 1988a. "Approaches to the Teaching of STS in Science Education." *International Journal of Science Education* 10(4): 346–356.

———. 1988b. "Science for All—A Vision Splendid." *Proceedings of Royal Society of New Zealand* 116: 191–197.

———. 1990. "What Will Science Education Do About Technology?" *The Australian Science Teachers Journal* 36(3): 8–21.

———. 1991. "Science and Technology." In *Handbook of Research on Curriculum*, ed. P. Jackson [for American Education Research Association]. New York: Macmillan.

Ferster, C. S., and B. F. Skinner. 1957. *Schedules of Reinforcement*. New York: Appleton-Century-Crofts.

Feyerabend, P. 1975. *Against Method*. New York: Schocken.

Fitzpatrick, F., ed. 1960. *Policies for Science Education*. New York: Teachers College Press, Columbia University.

Fitzpatrick, F., and Fensham, P. 1988. "Approaches to the Teaching of STS in Science Education." *International Journal of Science Education* 10(4): 346–356.

Fleming, R. 1989. "Literacy for a Technological Age." *Science Education* 73(4): 391–404.

Fourez, G. 1989. "Scientific Literacy, Societal Choices, and Ideologies." In *Scientific Literacy*, ed. A. B. Champagne, B. E. Lovitts, and B. Callinger. Washington, DC: American Association for the Advancement of Science.

Fuhrman, S. H., ed. 1993. *Designing Coherent Education Policy: Improving the System*. San Francisco: Jossey-Bass.

Fullan, M. G. 1982. *The Meaning of Educational Change*. New York: Teachers College Press, Columbia University.

———. 1993. *Change Forces: Probing the Depths of Education Reform*. New York: The Falmer Press.

———. 1996. "Turning Systematic Thinking on Its Head." *Phi Delta Kappan* 77(6): 420–423.

Fullan, M. G., and M. B. Miles. 1992. "Getting Reform Right: What Works and What Doesn't." *Phi Delta Kappan* 745–752.

Fullan, M. G., and S. Steigelbauer. 1991. *The New Meaning of Educational Change,* 2d ed. New York: Teachers College Press, Columbia University.

Gagne, R. M., and L. T. Brown. 1961. "Some Factors in the Programming of Conceptual Learning." *Journal of Experimental Psychology* 62:313–321.

Gardner, J. 1990. *On Leadership.* New York: Free Press.

Garrison, J. W., and M. Bentley. 1990. "Teaching Scientific Method: The Topic of Confirmation and Falsification." *School Science and Mathematics* 90: 180–197.

Gatewood, C. 1968. "The Science Curriculum Viewed Nationally." *The Science Teacher* 35: 20.

Gibbons, M. 1976. *The New Secondary Education.* Bloomington, IN: Phi Delta Kappa.

Glatthorn, A. 1987. *Curriculum Renewal.* Alexandria, VA: Association for Supervision and Curriculum Development.

Gould, S. J. 1987. *Time's Arrow, Time's Cycle.* Cambridge, MA: Harvard University Press.

Graff, H. 1986. "The Legacies of Literacy: Continuities and Contradictions in Western Society and Culture." In *Literacy, Society, and Schooling,* ed. S. deCastell, A. Luke, and K. Egan. Cambridge: Cambridge University Press.

Graubard, S. R. 1983. "Nothing to Fear, Much to Do." *Daedalus: Journal of the American Academy of Arts and Sciences* 112(2): 231–248.

Graubard, S. R., ed. 1983. "Scientific Literacy." *Daedalus: Journal of the American Academy of Arts and Sciences* 112(2).

Grobman, A. B. 1969. *The Changing Classroom: The Role of the Biological Sciences Curriculum Study.* Garden City, NY: Doubleday.

Gross, P., and N. Levitt. 1994. *Higher Superstition: The Academic Left and Its Quarrels with Science.* Baltimore, MD: Johns Hopkins University Press.

Gutherie, L. F., and C. Leventhal. nd. *Opportunities for Scientific Literacy for High School Students.* Far West Laboratory for Educational Research and Development. EDC263017.

Hall, G. E. 1989. "Changing Practice in High Schools: A Process, Not an Event." In *High School Biology: Today and Tomorrow,* ed. N. Grossblatt, 298–323. Washington, DC: National Academy Press.

———. 1992. "The Local Educational Change Process and Policy Implementation." *Journal of Research in Science Teaching* 29(8): 877–904.

Hall, G. E., and S. M. Hord. 1987. *Change in Schools: Facilitating the Process.* Albany: State University of New York Press.

Hall, G. E., and S. F. Loucks. 1978. "Teacher Concerns as a Basis for Facilitating and Personalizing Staff Development." *Teachers College Record* 80(1): 36–53.

Hardin, G. 1968. "The Tragedy of the Commons." *Science* 162: 1243.

Harms, N. 1977. *Project Synthesis: An Interpretive Consolidation of Research Identifying Needs in Natural Sciences Education.* A proposal prepared for the National Science Foundation, University of Colorado, Boulder, CO.

Harms, N., and S. Kahl. 1980. *Project Synthesis: Final Report.* Submitted to the National Science Foundation, University of Colorado, Boulder, CO.

Harms, N., and R. E. Yager. eds. 1981. *What Research Says to the Science Teacher, Vol. 3.* Washington, DC: National Science Teachers Association.

Harrison, A. J. 1984. "Common Elements and Interconnections." *Science* 224(4652): 939–942.

Harvard Committee. 1945. *General Education in a Free Society.* Cambridge, MA: Harvard University Press.

Hawkins, D. 1965. "Messing About in Science." *Science and Children* 2(6): 5–9.

Hawkins, J., and R. Pea. 1987. "Tools for Bridging the Cultures of Everyday and Scientific Thinking." *Journal of Research in Science Teaching* 24: 291–307.

Hazen, R., and J. Trefil. 1991a. *Science Matters: Achieving Scientific Literacy.* New York: Doubleday.

———. 1991b. "General Science Courses Are the Key to Scientific Literacy." *The Chronicle of Higher Education.*

Helgeson, S. L., P. E. Blosser, and R. W. Howe. 1977. *The Status of Pre-College Science, Mathematics, and Social Science Education: 1955–1975.* Vol. 1. Columbus, OH: Center on Science and Mathematics Education, Ohio State University.

Hentoff, N. 1966. *Our Children Are Dying.* New York: Viking Press.

Herbart, J. 1901. *Outlines of Educational Doctrine,* trans. C. DeGarmo; ed. A. Lange. New York: Macmillan.

Hersher, L. 1988. "On the Absence of Revolution in Biology." *Perspectives in Biology and Medicine* 31: 318.

Hetherington, N. 1982. "The History of Science and the Teaching of Science Literacy." *Journal of Thought* 17: 53–66.

Hewson, M. G. A. 1984. "The Role of Conceptual Conflict in Conceptual Change and the Design of Science Instruction." *Instructional Science* 13:1–13.

———. 1986. "The Acquisition of Scientific Knowledge: Analysis and Representation of Student Conceptions Concerning Density." *Science Education* 70: 159–170.

Hickman, F. M., J. Patrick, and R. W. Bybee. 1987. *Science-Technology-Society: A Framework for Curriculum Reform in Secondary School Science and Social Studies.* Boulder, CO: Social Science Education Consortium.

Hirsh, E. D. 1987. *Cultural Literacy: What Every American Needs to Know.* Boston: Houghton Mifflin.

Hlebowith, P. S., and S. E. Hudson. 1991. "Science Education and the Reawakening of the General Education Idea." *Science Education* 75(5): 563–576.

Hodgkinson, H. 1991. "Reform Versus Reality." *Phi Delta Kappan* 73(1): 9–16.

Holt, J. C. 1982. *How Children Fail.* New York: Delacorte.

Holton, G. 1993. *Science and Anti-Science.* Cambridge, MA: Harvard University Press.

Holzman, M. 1993. "What Is Systemic Change?" *Educational Leadership* 51(1): 18.

House, P. 1986. "Now More Than Ever: The Alliance of Science and Mathematics." *School Science and Mathematics* 86(6): 456–460.

Hughes, T. 1976. "The Science-Technology Interaction: The Case of High-Voltage Power Transmission Systems." *Technology and Culture* 17: 646–662.

Hurd, P. D. 1958. "Science Literacy: Its Meaning for American Schools." *Educational Leadership* 16: 13–16.

———. 1962. *Biology Education in American Schools: 1890–1960*. Washington, DC: American Institute of Biological Scientists.

———. 1969. *New Directions in Teaching Secondary School Science*. Chicago: Rand McNally.

———. 1970. "Scientific Enlightenment for an Age of Science." *The Science Teacher* 37: 13–15.

———. 1975. "Science, Technology, and Society: New Goals for Interdisciplinary Science Teaching." *The Science Teacher* 42: 27–30.

———. 1987. "A Nation Reflects: The Modernization of Science Education." *Bulletin of Science, Technology, and Society* 7: 9–13.

———. 1990. "Historical and Philosophical Insights on Scientific Literacy." *Bulletin of Science, Technology, and Society* 10: 133–136.

———. 1992. "First in the World by 2000: What Does It Mean?" *Education Week* 12(2): 36, 28.

Hurd, P. D., R. W. Bybee, J. B. Kahle, and R. Yager. 1981. "Biology Education in Secondary Schools of the United States." *The American Biology Teacher* 42(7): 388–410.

Hurd, P. D., and J. J. Gallagher. 1966. "Goals Related to the Social Aspects of Science." In *Sequential Programs in Science for a Restructional Curriculum*, 12–18. Cleveland, OH: Educational Research Council.

Ilg, F., and L. Ames. 1955. *Child Behavior*. New York: Dell.

Jenkins, E. 1990a. "Scientific Literacy and School Science Education." *School Science Review* 71(256): 43–51.

———. 1990b. "The History of Science in British Schools: Retrospect and Prospect." *International Journal of Science Education* 12(3): 274–281.

———. 1992. "School Science Education: Toward a Reconstruction." *Journal of Curriculum Studies* 24(3): 229–246.

Johnson, D. W., and R. T. Johnson. 1987. *Learning Together and Alone*, 2d. ed. Englewood Cliffs, NJ: Prentice Hall.

Johnson, D. W., R. T. Johnson, and E. J. Holubec. 1986. *Circles of Learning: Cooperation in the Classroom*. Edina, MN: Interaction Books.

Johnson, D. W., R. T. Johnson, and G. Maruyama. 1983. "Interdependence and Interpersonal Attraction Among Heterogeneous and Homogeneous Individuals: A Theoretical Formulation and a Meta-Analysis of the Research." *Review of Educational Research* 52: 5–54.

Johnson, J. R. 1989. *Technology*. Washington, DC: American Association for the Advancement of Science.

Johnson, P. G. 1962. "The Goals of Science Education." *Theory into Practice* 1: 239–244.

Joshua, S., and J. J. Dupin. 1987. "Taking into Account Student Conceptions in Instructional Strategy: An Example in Physics." *Cognition and Instruction* 42(2):117–135.

Karplus, R. 1964. "The Science Curriculum Improvement Study." *Journal of Research in Science Teaching* 2: 293–303.

———. 1971. "Some Thoughts on Science Curriculum Development." In *Con-

fronting Curriculum Reform, ed. E. W. Eisner. Boston: Little, Brown and Company, Inc.

Klopfer, L. 1969. "The Teaching of Science and the History of Science." *Journal of Research in Science Teaching* 6: 87–95.

———. 1976. "Scientific Literacy Reexamined." *Science Education* 60(1): 95.

Koelsche, C. 1965. "Scientific Literacy as Related to the Media of Mass Communications." *School Science and Mathematics* 65: 719–724.

Kohl, H. R. 1967. *36 Children.* New York: New American Library.

Kondo, A. 1972. "Scientific Literacy: A View from a Developing Country." *NASSP Bulletin* 56(360): 28–37.

Kozol, J. 1967. *Death at an Early Age: The Destruction of the Hearts and Minds of Negro Children in the Boston Public Schools.* Boston: Houghton Mifflin.

Kuhn, D. 1992. "Thinking as Argument." *Harvard Educational Review* 62(2): 155–178.

Kuhn, T. S. 1962. *The Structure of Scientific Revolutions.* Princeton, NJ: Princeton University Press.

———. 1977. *The Essential Tension: Selected Studies in Scientific Tradition and Change.* Chicago: University of Chicago Press.

Kusch, P. 1960. "Educating for Scientific Literacy in Physics." *School and Society* 80(2173): 198–201.

Kyle, W. C. 1980. "The Distinction Between Inquiry and Scientific Inquiry and Why High School Students Should Be Cognizant of the Distinction." *Journal of Research in Science Teaching* 17(2).

———. 1991. "The Reform Agenda and Science Education: Hegemonic Control Vs. Counterhegemony." *Science Education* 75(4): 403–412.

Kyle, W. C., S. Abell, and J. Shymansky. 1989. "Enhancing Prospective Teachers' Conceptions of Teaching and Science." *Journal of Science Teacher Education* 1(3): 10–13.

Lakatos, I., and A. Musgrave. 1970. *Criticism and the Growth of Knowledge.* Cambridge: Cambridge University Press.

Laszlo, E. 1972. *A Strategy for the Future.* New York: George Braziller.

Lavach, J. 1969. "Organization and Evaluation of an In-service Program in the History of Science." *Journal of Research in Science Teaching* 6: 166–170.

Lawson, A. E. 1988. "A Better Way to Teach Biology." *The American Biology Teacher* 50(5): 266–278.

Lawson, A., M. Abraham, and J. Renner. 1989. A Theory of Instruction: Using the Learning Cycle to Teach Science Concepts and Thinking Skills. National Association for Research in Science Teaching. Monograph One. (Contact A. Lawson, Arizona State University, Tempe.)

Lawson, A. E., and W. T. Wollman. 1975. "Physics Problems and the Process of Self-Regulation." *The Physics Teacher* 13(8): 470–475.

Lawton, M. 1996. "Summit Accord Calls for Focus on Standards." *Education Week* XV(28): 1; 14–15.

Layton, D. 1973. "The Secondary School Curriculum and Science Education." *Physics Education* 8(3): 19–23.

———. 1988. "Revaluating the T in STS." *International Journal of Science Education* 10(4): 367–378.

———. 1990. *Inarticulate Science?* Department of Education Lecture. Occasional Papers No. 17. Liverpool, UK: University of Liverpool.

———. 1991. "Science Education and Proxis: The Relationship of School Science to Practical Action." *Studies in Science Education* 19: 43–79.

———. 1992. "Values in Design and Technology." In *Make the Future Work: Appropriate Technology,* ed. C. Budgett-Meakin. New York: Longman.

Layton, D., D. Jenkins, and M. Donnelly. 1993. *Scientific and Technological Literacy, Meanings and Rationale: An Annotated Bibliography.* Leeds, UK: University of Leeds.

Lederman, N. 1986. "Students' and Teachers' Understanding of the Nature of Science: A Reassessment." *School Science and Mathematics* 86: 91–99.

Lederman, N., and M. O'Malley. 1990. "Students' Perception of Tentativeness in Science: Development, Use, and Sources of Change." *Science Education* 74: 2.

LeMahieu, P. A., and H. K. Foss. 1994. "Assumptions of Standards Based Reform and Their Implication for Policy and Practice." *The School Administrator* 51(5): 16–22.

Linn, M. 1986. *Establishing a Research Base for Science Education: Challenges, Trends and Recommendations.* Berkeley: University of California.

Little, J. W. 1993. "Teachers' Professional Development in a Climate of Educational Reform." *Educational Evaluation and Policy Analysis* 15: 129–151.

Loucks-Horsley, S. 1995. "Professional Development and the Learner Centered School." *Theory into Practice* 34(4): 265–271.

———. 1996. "Professional Development for Science Education: A Critical and Immediate Challenge." In *National Standards and the Science Curriculum: Challenges, Opportunities, and Recommendations,* ed. R. W. Bybee, 83–95. Dubuque, IA: Kendall/Hunt.

Loucks-Horsley, S., and S. Stiegelbauer. 1991. "Using Knowledge of Change to Guide Staff Development." In *Staff Development for Education in the 90s: New Demands, New Realities, New Perspectives,* ed. A. Lieberman and L. Miller. New York: Teachers College Press, Columbia University.

Mackay, L. 1971. "Development of Understanding About the Nature of Science." *Journal of Research in Science Teaching* 8: 57–66.

Martin, J. 1974. *The Education of Adolescents.* Washington, DC: U.S. Office of Education.

Martin, M. 1972. *Concepts of Science Education: A Philosophical View.* Greenview, IL: Scott Foresman.

May, R. 1991. *The Cry for Myth.* New York: Delta.

Mayer, W. V. 1976. "The BSCS Process of Curriculum Development." *Biological Sciences Curriculum Study Newsletter* 64, September: 4–10.

McGilly, K., ed. 1994. *Classroom Lessons: Integrating Cognitive Theory and Classroom Practice.* Cambridge, MA: MIT Press.

McInerney, J. D. 1986/1987. "Curriculum Development at the Biological Sciences Curriculum Study." *Educational Leadership,* Dec./Jan., 24–28.

McLaughlin, M., L. Shepard, and J. O'Day. 1995. *Improving Education Through Standards-Based Reform.* Stanford, CA: The National Academy of Education.

McQuigg, R. B. 1972. "Science and General Education." *NASSP Bulletin* 56(360): 38–41.

Mead, M., and M. Metraux. 1957. "Image of the Scientist Among High School Students." *Science* 126: 384–390.

Meadows, D., D. Meadows, J. Randers, and W. Behrens. 1972. *The Limits to Growth.* New York: Universe.

Miller, J. D. 1983a. *The American People and Science Policy.* New York: Pergamon Press.

———. 1983b. "Scientific Literacy: A Conceptual and Empirical Review." *Daedalus: Journal of the American Academy of Arts and Sciences* 112(2): 19–48.

Milson, J., and S. Bell. 1986. "Enhancement of Learning Through Integrating Science and Mathematics." *School Science and Mathematics* 86(6): 489–493.

Minstrell, J. A. 1984. "Teaching for the Development of Understanding of Ideas: Focus on Moving Objects." In *Observing Science Classrooms: Observing Science Perspectives from Research and Practice, 1984 AETS Yearbook.* Columbus, OH: Ohio State University.

———. 1989. "Teaching Science for Understanding." In *Toward the Thinking Curriculum: Current Cognitive Research, 1989 Yearbook of the ASCD,* ed. L. Resnick and L. Klopfer. Alexandria, VA: Association for Supervision and Curriculum Development.

Mitman, A. L., J. R. Mergendoller, V. A. Marchman, and M. J. Packer. 1987. "Instruction Addressing the Components of Scientific Literacy and Its Relation to Student Outcomes." *American Educational Research Journal* 24(4): 611–633.

Moore, J. A. 1983. *Science as a Way of Knowing.* Vol. 1: *Evolutionary Biology.* Baltimore, MD: American Society of Zoologists.

———. 1984. *Science as a Way of Knowing.* Vol. 2: *Human Ecology.* Baltimore, MD: American Society of Zoologists.

———. 1985. *Science as a Way of Knowing.* Vol. 3: *Genetics.* Baltimore, MD: American Society of Zoologists.

———. 1986. *Science as a Way of Knowing.* Vol. 4: *Developmental Biology.* Baltimore, MD: American Society of Zoologists.

———. 1987. *Science as a Way of Knowing.* Vol. 5: *Form and Function.* Baltimore, MD: American Society of Zoologists.

———. 1988a. *Science as a Way of Knowing.* Vol. 6: *Cell and Molecular Biology.* Baltimore, MD: American Society of Zoologists.

———. 1988b. "Teaching the Sciences as Liberal Arts—Which, of Course, They Are." *Journal of College Teaching* 17: 444–451.

———. 1989. *A Conceptual Framework for Biology.* Parts I and II. Baltimore, MD: American Society of Zoologists.

———. 1990. *A Conceptual Framework for Biology:* Part III. Baltimore, MD: American Society of Zoologists.

Mullis, I., and L. Jenkins. 1988. *The Science Report Card: Elements of Risk and Recovery.* Princeton, NJ: Educational Testing Service.

Murnane, R., and S. Raizen, eds. 1988. *Improving Indicators of the Quality of Science and Mathematics Education in Grades K-12*. Washington, DC: National Academy Press.

National Assessment of Educational Progress. 1978a. *Science Achievement in the Schools: A Summary of Results from the 1976–77 National Assessment of Science*. Denver, CO: Education Commission of the States.

———. 1978b. *Three National Assessments of Science: Changes in Achievement, 1969–1977*. Denver, CO: Education Commission of the States.

National Association of Secondary School Principals. 1972. *NASSP Bulletin* 56(360).

National Center for Educational Statistics. 1996. *Pursuing Excellence: A Study of U.S. Eighth Grade Mathematics and Science Teaching, Learning, Curriculum, and Achievement in International Context*. Washington, DC: U.S. Government Printing Office.

National Center for History in the Schools. 1996. *National Standards for History*. Los Angeles, CA: National Center for History in the Schools.

National Commission on Excellence in Education. 1983. *A Nation at Risk: The Imperative for Educational Reform*. Washington, DC: U.S. Department of Education.

National Council for the Social Studies. 1983. "Guidelines for Teaching Science-Related Social Issues." *Social Education* 258–261.

National Council of Teachers of Mathematics. 1989. *Curriculum and Evaluation Standards for School Mathematics*. Reston, VA: National Council of Teachers of Mathematics.

National Education Goals Panel. 1994. *The National Education Goals Report: Building a Nation of Learners*. Washington, DC: U.S. Government Printing Office.

National Research Council. 1989. *Everybody Counts: A Report to the Nation on the Future of Mathematics Education*. Washington, DC: National Academy Press.

———. 1990. *Fulfilling the Promise: Biology Education in the Nation's Schools*. Washington, DC: National Academy Press.

———. 1992. Coordinating Council for Education. *Science Framework Summaries*. Washington, DC: National Research Council.

———. 1994. *National Science Education Standards (Draft)*. Washington, DC: National Academy Press.

———. 1995. *National Science Education Standards (Draft for Review Comments)*. Washington, DC: National Academy Press.

———. 1996a. *National Science Education Standards*. Washington, DC: National Academy Press.

———. 1996b. *What Can We Learn?* Washington, DC: National Academy Press.

National Science Board. 1983. *Educating Americans for the 21st Century*. Washington, DC: U.S. Government Printing Office.

National Science Foundation. 1996. *The Learning Curve: What We Are Discovering About U.S. Science and Mathematics Education*. Arlington, VA: National Science Foundation.

National Science Teachers Association. 1964. *Theory into Action*. Washington, DC: National Science Teachers Association.

———. 1971. *School Science Education for the 1970s: An NSTA Position Statement.* Washington, DC: National Science Teachers Association.

———. 1982. *Science Education for the 1980s. Science-Technology-Society: An NSTA Position Statement.* Washington, DC: National Science Teachers Association.

———. 1990. *Science Teachers Speak Out: The NSTA Lead Paper on Science and Technology Education for the 21st Century.* Washington, DC: National Science Teachers Association.

———. 1992. *Scope, Sequence, and Coordination of Secondary School Science. The Content Core: A Guide for Curriculum Designers.* Washington, DC: National Science Teachers Association.

National Society for the Study of Education. 1947. *Science Education in American Schools: Forty-Sixth Yearbook of the National Society for the Study of Education.* Chicago: University of Chicago Press.

Nodine, C., J. Gallagher, and D. Humphreys, eds. 1972. *Piaget and Inhelder on Equilibration.* Philadelphia: The Jean Piaget Society, Temple University.

Norman, C. 1981. *The God That Limps: Science and Technology in the Eighties.* New York: W. W. Norton.

Novak, J. D. 1987. *Proceedings of the Second International Seminar on Misconceptions and Educational Strategies in Science and Mathematics.* New York: Columbia University Press.

Novak, J. D., and D. B. Gowin. 1984. *Learning How to Learn.* Cambridge: Cambridge University Press.

Numbers, R. 1982. "The History of American Medicine: A Field in Ferment," in "The Promise of American History. Progress and Prospects (A Special Edition of Reviews)." *American History* 10: 245–263.

Oakes, J. 1990a. *Lost Talent: The Underrepresentation of Women, Minorities, and Disabled Persons in Science.* Santa Monica, CA: Rand Corporation.

———. 1990b. *Multiplying Inequalities and the Effects of Race, Social Class, and Teaching on Opportunities to Learn Mathematics and Science.* Santa Monica, CA: Rand Corporation.

O'Hearn, G. T. 1972. "Environmental Education: The New Scientific Literacy." *NASSP Bulletin* 56(360): 21–27.

———. 1976. "Scientific Literacy and Alternative Futures." *Science Education* 60(1): 103–114.

O'Neil, J. 1993. "Turning the System on Its Head." *Educational Leadership* 51(1): 8–13.

Osborne, R., and P. Freyberg. 1985. *Learning in Science: The Implications of Children's Science.* Portsmouth, NH: Heinemann.

Patrick, J., and R. C. Remy. 1985. *Connecting Science, Technology, and Society in the Education of Citizens.* Boulder, CO: Social Science Education Consortium.

Patton, J. B. 1972. "Geology: A Necessary Discipline for Ecological Problem Solving." *NASSP Bulletin* 56(360): 13–20.

Pauly, E. 1992. "Classrooms Matter More than Policies." *Education Week* (May): 36.

Pella, M. O. 1967. "Science Literacy and the High School Curriculum." *School Science and Mathematics* 67: 346–356.

————. 1976. "The Place and Function of Science for a Literate Citizenry." *Science Education* 60(1).

Pella, M. O., G. T. O'Hearn, and C. W. Gale. 1966. "Referents to Scientific Literacy." *Journal of Research in Science Teaching* 4: 199–208.

Piaget, J., 1975. *The Development of Thought.* New York: Viking Press.

Piaget, J., and B. Inhelder. 1969. *The Psychology of the Child.* New York: Basic Books.

Posner, G. J. 1992. *Analyzing the Curriculum.* New York: McGraw Hill.

Price, D. K. 1965. *The Scientific Estate.* Cambridge, MA: Harvard University Press.

Progressive Education Association. 1938. *Science in General Education.* New York: Appleton-Century-Crofts.

Raizen, S. 1991. "The Reform of Science Education in the U.S.A.: Deja Vu or de Nova?" *Studies in Science Education* 19: 1–41.

Rapoport, A. 1968. "Foreword." In *Modern Systems Research for the Behavioral Scientist,* ed. W. F. Buckley. Chicago: Aldine Press.

Ravitch, D. 1983. *The Troubled Crusade: American Education 1945–1980.* New York: Basic.

————. 1995. *National Standards in American Education.* Washington, DC: The Brookings Institute.

Ravitch, D., and C. E. Finn, Jr. 1987. *What Do Our 17-year Olds Know?* New York: Harper and Row.

Renner, J. W., M. R. Abraham, and H. H. Bernie. 1985. "The Importance of the Form of Student Acquisition of Data in Physics Learning Cycles." *Journal of Research in Science Teaching* 22:303–325.

————. 1988. "The Necessity of Each Phase of the Learning Cycle in Teaching High School Physics." *Journal of Research in Science Teaching* 25(1): 39–58.

Renner, J. W., and A. E. Lawson. 1973. "Promoting Intellectual Development Through Science Teaching." *The Physics Teacher* 11(5): 273–276.

Rhodes, S. L. 1992. "Climate Change Management Strategies: Lessons from a Theory of Large-Scale Policy." *Global Environmental Change* 2(3): 205–213.

Roberts, D. 1980. "Theory, Curriculum Development, and the Unique Events of Practice." In *Seeing Curriculum in a New Light,* ed. H. Munby, G. Orpwood, and T. Russell, 65–87. Ontario, Canada: OISE Press, The Ontario Institute for Studies in Education.

————. 1982. "Developing the Concept of Curriculum Emphasis in Science Education." *Science Education* 66(2): 243–260.

————. 1983. *Scientific Literacy: Towards Balance in Setting Goals for School Science Programs.* Ottawa, Canada: Science Council of Canada.

————. 1988. "What Counts as Science Education?" In *Development and Dilemmas in Science Education,* ed. P. Fensham. New York: Falmer Press.

————. 1995. "Scientific Literacy: The Importance of Multiple Curriculum Emphases." In *Redesigning the Science Curriculum: A Report on the Implications of Standards and Benchmarks for Science Education,* ed. R. W. Bybee and J. D. McInerney, 75–80. Colorado Springs, CO: Biological Sciences Curriculum Study.

Robinson, J. 1968. *The Nature of Science and Science Teaching.* Belmont, CA: Wadsworth.

Roseman, J. E. 1996. "Implementing Specific Learning Goals: Lessons from Project 2061." In *National Standards & the Science Curriculum: Challenges, Opportunities, & Recommendations,* ed. R. W. Bybee, 55–74. Dubuque, IA: Kendall/Hunt.

Roth, K. J., and C. W. Anderson. 1988. "Promoting Conceptual Change Learning from Science Textbooks." In *Improving Learning: New Perspectives,* ed. P. Ramsden. New York, NY: Kogan Page.

Rothman, R. 1995. *Measuring Up: Standards, Assessment, and School Reform.* San Francisco: Jossey-Bass.

Roughead, W. G., and J. M. Scandura. 1968. "What Is Learned in Mathematical Discovery?" *Journal of Educational Psychology* 59:283–289.

Rubba, P., J. Horner, and J. Smith. 1981. "A Study of Two Misconceptions About the Nature of Science Among Junior High School Students." *School Science and Mathematics* 81: 221–226.

Russell, B. 1951. *The Impact of Science on Society.* New York: Columbia University Press.

Rutherford, F. J. 1964. "The Role of Inquiry in Science Teaching." *Journal of Research in Science Teaching* 2: 80–84.

———. 1972. "A Humanistic Approach to Science Teaching." *NASSP Bulletin* 53(361): 53–63.

———. 1992. "On Being Inclusive." *2061 Today* 2(2): 5.

Rutherford, F. J., and A. Ahlgren. 1989. *Science for All Americans: A Project 2061 Report.* Washington, DC: American Association for the Advancement of Science.

Rutherford, F. J., G. Holton, and F. Watson. 1970. *The Project Physics Course: Text.* New York: Holt, Rinehart, and Winston.

Salk, J., and J. Salk. 1981. *World Population and Human Values.* New York: Harper and Row.

Sarason, S. B. 1991. *The Predictable Failure of Education Reform.* San Francisco: Jossey-Bass.

Saylor, G., and W. Alexander. 1974. *Planning Curriculum for Schools.* New York: Holt, Rinehart, and Winston.

Scheffler, I. 1960. *The Language of Education.* Springfield, IL: Charles C Thomas.

Schlesinger, A. M. 1992. *The Disuniting of America.* New York: W. W. Norton.

Schmidt, W. H., and C. C. McKnight. 1995. "Surveying Educational Opportunity in Mathematics and Science: An International Perspective." *Educational Evaluation and Policy Analysis* 17(3): 337–353.

Schmidt, W., C. McKnight, S. Raizen. 1997. *Splintered Vision: An Investigation of U.S. Science and Mathematics Education.* Executive Summary. East Lansing, MI: Michigan State University, U.S. National Research Center for the Third International Mathematics and Science Study.

Schubert, W. H. 1993. "Curriculum Reform." In *Challenges and Achievements of American Education, 1993 Yearbook of the ASCD,* ed. G. Cavelti. Alexandria, VA: Association for Supervision and Curriculum Development.

Schulman, P. R. 1980. *Large-Scale Policy Making.* New York: Elsevier-North Holland.

Schwab, J. 1960. "What Do Scientists Do?" *Behavioral Science* 5: 1.

————. 1966. *The Teaching of Science.* Cambridge, MA: Harvard University Press.

————. 1970. *The Practical: A Language for Curriculum.* Washington, DC: National Education Association.

Scientific Literacy Group. 1987. *Scientific Literacy Papers.* Oxford: University of Oxford.

Scriven, J. 1973. *The Methodology of Evaluation. Education Evaluation: Theory and Practice.* Belmont, CA: Wadsworth.

Senge, P. 1990a. *The Fifth Discipline.* New York: Doubleday Currency.

————. 1990b. "The Leader's New Work: Building Learning Organizations." *Sloan Management Review* 32(1): 7–23.

Shamos, M. 1963. "The Price of Scientific Literacy." *National Association of School Principals* 47: 41–51.

————. 1984. "You Can Lead a Horse to Water . . ." *Educational Leadership* 20–23.

————. 1988a. "A False Alarm in Science Education." *Issues in Science and Technology* IV(3): 65–69.

————. 1988b. "The Lesson Every Child Need Not Learn." *The Sciences* 14–20.

————. 1989. "Views of Scientific Literacy in Elementary School Science Programs: Past, Present, and Future." In *This Year in School Science 1989: Scientific Literacy,* ed. A. B. Champagne and B. E. Lovitts. Washington, DC: American Association for the Advancement of Science.

————. 1990. "Science Literacy Where It Counts." *Journal of College Science Teaching* 88–89.

————. 1995. *The Myth of Scientific Literacy.* New Brunswick, NJ: Rutgers University Press.

Shen, B. S. P. 1975. "Science Literacy: The Public Need." *The Sciences* Jan.–Feb: 27–29.

Showalter, V. 1974. "What Is Unified Science Education? Program Objectives and Scientific Literacy." *Prism* II-2: 1–6.

Shulman, L. 1986. "Those Who Understand: Knowledge Growth in Teaching." *Educational Researcher* 15: 4.

Shymansky, J. A., W. C. Kyle, and J. M. Alport. 1983. "The Effects of New Science Curricula on Student Performance." *Journal of Research in Science Teaching* 20(5): 387–404.

Silberman, C. E. 1970. *Crisis in the Classroom: The Remaking of American Education.* New York: Random House.

Silver, C. S., and R. DeFries. 1990. *One Earth, One Future.* Washington, DC: National Academy Press.

Sitarz, D., ed. 1993. *Agenda 21: The Earth Summit Strategy to Save Our Planet.* Boulder, CO: Earthpress.

Smith, C., S. Carey, and M. Wiser. 1985. "On Differentiation: A Case Study of the Development of the Concepts of Size, Weight, and Density." *Cognition* 21(3): 177–237.

Smith, M., and J. O'Day. 1991. "Systemic School Reform." In *The Politics of Curriculum and Testing,* ed. S. Fuhrman and B. Malen. Bristol, PA: The Falmer Press.

Smith, N. 1974. "The Challenge of Scientific Literacy." *The Science Teacher* 41: 34–35.

Snow, C. P. 1962. *The Two Cultures and the Scientific Revolution*. New York: Cambridge University Press.

Sonneborn, T. M. 1972. "Secondary School Preparation for Making Biological Decisions." *NASSP Bulletin* 56(360): 1–12.

Spector, B., and N. Lederman. 1990. *Science and Technology: A Human Enterprise*. Dubuque, IA: Kendall/Hunt.

SRI International. 1995a. *Evaluation of the Dwight D. Eisenhower Mathematics and Science Education State Curriculum Frameworks Projects: First Report (Draft)*. Menlo Park, CA: SRI.

———. 1995b. *Evaluation of the Dwight D. Eisenhower Mathematics and Science Regional Consortiums Program: First Interim Report (Draft)*. Menlo Park, CA: SRI.

Stake, R. E., and J. A. Easley. 1978. *Case Studies in Science Education*. Urbana, IL: Center for Institutional Research and Curriculum Evaluation, University of Illinois.

Stevenson, H., and J. Stigler. 1992. *The Learning Gap*. New York, NY: Touchstone.

Sykes, G., and P. Plastrik. 1993. *Standard Setting as Educational Reform*. Trends and Issues Paper No. 8. Washington, DC: ERIC Clearinghouse on Teacher Education and American Association of Colleges for Teacher Education.

Tanner, D., and L. Tanner. 1990. *History of the School Curriculum*. New York: Macmillan.

Task Force on Women, Minorities, and the Handicapped in Science and Technology. 1988. *Changing America: The New Face of Science and Engineering, Interim Report*. Washington, DC: The Task Force.

Thackray, A. 1980. "History of Science." In *A Guide to the Culture of Science Technology and Medicine,* ed. P. T. Durbin. New York: Free Press.

Thorsheim, H. 1986. "Systems Thinking: The Positive Influence of STS on Educational Motivation." In *Science-Technology-Society NSTA Yearbook: 1985,* ed. R. W. Bybee. Washington, DC: National Science Teachers Association.

Tisher, R. P., ed. 1985. *Research in Science Education.* Vol. 15. Selection of papers from the annual conference of the Australian Science Education Research Association, Rockhampton, Queensland, Australia. ERIC Document Reproduction Service No. ED 267 974.

Toulmin, S. E. 1972. *Human Understanding*. Princeton, NJ: Princeton University Press.

———. 1982. "The Construal of Reality: Criticism in Modern and Postmodern Science." *Critical Inquiry* 9: 93.

Tressel, G. W. 1994. "Thirty Years of 'Improvement' in Precollege Math and Science Education." *Journal of Science Education and Technology* 3(2): 77–88.

Tyack, D., and L. Cuban. 1995. *Tinkering Toward Utopia: A Century of Public School Reform*. Cambridge: Harvard University Press.

Tyler, R. 1949. *Basic Principles of Curriculum and Instruction*. Chicago: The University of Chicago Press.

U.S. Congress Office of Technology Assessment. 1988a. *Grade School to Grad School*. OTA-SET-377. Washington, DC: U.S. Government Printing Office.

———. 1988b. *Elementary and Secondary Education for Science and Engineering—A Technical Memorandum.* OTA-TM-SET-41. Washington, DC: U.S. Government Printing Office.

U.S. National Research Center for the Third International Mathematics and Science Study. 1997. *Splintered Vision: An Investigation of U.S. Science and Mathematics Education. Executive Summary.* Lansing, MI: Michigan State University.

Vinovskis, M. A. 1996. "An Analysis of the Concept and Uses of Systemic Educational Reform." *American Educational Research Journal* 33(1): 53–85.

Vygotsky, L. S. 1962. *Thought and Language.* Cambridge, MA: MIT Press.

———. 1978. *Mind in Society: The Development of Higher Psychological Processes.* Cambridge, MA: Harvard University Press.

Wagner, P. A. 1983. "The Nature of Paradigmatic Shifts and the Goals of Science Education." *Science Education* 67: 605–613.

Ward, B., and R. Dubos. 1972. *Only One Earth: The Care and Maintenance of a Small Planet.* New York: W. W. Norton.

Watson, F. 1967. "Why Do We Need More Physics Courses?" *The Physics Teacher* 5(5): 13–18.

Weiss, I. 1978. *Report of the 1977 National Survey of Science, Mathematics, and Social Studies Education.* Washington, DC: U.S. Government Printing Office.

———. 1987. *Report of the 1985–86 National Survey of Science and Mathematics Education.* Triangle Park, NC: Research Triangle Institute.

Weisskopf, V. 1989. *The Privilege of Being a Physicist.* New York: W. H. Freeman.

Welch, W., L. Klopfer, G. Aikenhead, and J. Robinson. 1981. "The Role of Inquiry in Science Education: Analysis and Recommendations." *Science Education* 65: 33–50.

White, R. T. 1985. "Interview Protocols and Dimensions of Cognitive Structure." In *Cognitive Structure and Conceptual Change,* ed. L. H. T. West and A. L. Pines. Orlando, FL: Academic Press.

White, R., and R. Gunstone. 1992. *Probing Understanding.* Bristol, PA: The Falmer Press.

Wilson, K. G., and B. Davis. 1994. *Redesigning Education.* New York: Henry Holt.

Wittlin, A. 1963. "Scientific Literacy Begins in the Elementary School." *Science Education* 47(4): 331–342.

Wood, R., M. O. Pella, and G. O'Hearn. 1968. "Scientifically and/or Technologically Oriented Articles in Selected Newspapers." *Journal of Research in Science Teaching* 5: 151–153.

Yager, R. E. 1984. "Science and Technology in General Education." In *NSTA 1984 Yearbook,* ed. R. W. Bybee, J. Carlson, and A. J. McCormack. Washington, DC: National Science Teachers Association.

Yager, R. E., ed. 1993. *What Research Says to the Science Teacher.* Vol. 7: *The Science-Technology-Society Movement.* Arlington, VA: National Science Teachers Association.

Zeidler, D. L., and N. G. Lederman. 1989. "The Effects of Teachers' Language on Students' Conceptions of the Nature of Science." *Journal of Research in Science Teaching* 2(26): 771–783.

INDEX